ARAB PETRO- POLITICS

ABDULAZIZ AL-SOWAYEGH

ST. MARTIN'S PRESS
New York

© 1984 Abdulaziz al-Sowayegh
All rights reserved. For information, write:
St. Martin's Press, Inc., 175 Fifth Avenue, New York, NY 10010
Printed in Great Britain
First published in the United States of America in 1984

Library of Congress Cataloging in Publication Data

al-Sowayegh, Abdulaziz H.
    Arab petropolitics.

    Includes index.
    1. Petroleum industry and trade—Political aspects—
Arab countries.   2. Arab countries—Politics and government
—1945-   . 3. Jewish-Arab relations—Economic aspects.
I. Title.
HD9578.A55A54829   1984   327.1'11'09174927   83-40196
ISBN 0-312-04718-5

Printed and bound in Great Britain

CONTENTS

TABLES

FIGURES

ABBREVIATIONS

AIOC	Anglo-Iranian Oil Company
AOEC	Arab Oil Exports Committee
APOC	Anglo-Persian Oil Company
CFP	Compagnie Française des Pétroles
EEC	European Economic Community
ENI	Ente Nazionale Idrocarburi
ERAP	Enterprise de Recherches et d'Activités Pétrolières
IBRD	International Bank for Reconstruction and Development
IMF	International Monetary Fund
INOC	Iraq National Oil Company
IPC	Iraq Petroleum Company
KNPC	Kuwait National Petroleum Company
LINOCO	Libyan National Oil Company
MDECT	Mutual Defense and Economic Co-operation Treaty
MEDO	Middle East Defense Organization
NATO	North Atlantic Treaty Organization
NBC	National Broadcasting Corporation
NIOC	National Iranian Oil Company
OAPEC	Organization of Arab Petroleum Exporting Countries
OECD	Organization for Economic Co-operation and Development
OIDC	Oil-importing developing country
OPEC	Organization of Petroleum Exporting Countries
PLO	Palestine Liberation Organization
TPC	Turkish Petroleum Company
UAE	United Arab Emirates
UN	United Nations
UNSCOP	United Nations Special Committee on Palestine
US	United States
USSR	Union of Soviet Socialist Republics
WZO	World Zionist Organization

PREFACE

Without oil, the Middle East would be very different indeed. Oil has changed the political and economic structures and policies of the Middle East, and dramatically influenced political alignments, both within the region and between the region and the world's great powers. The world powers, in their turn, have found their oil interests and ambitions in the Middle East intruding increasingly into their relationships with each other, thus launching the region into the center stage of international tension.

Energy plays a pivotal role in the world economy today, and oil provides more than half of world energy consumption. Oil has become the primary fuel of the industrialized countries. In 1960, for example, oil supplied about one third of Western Europe's energy requirements; by 1980 this figure had risen to 60 percent. Japan depends upon oil for 75 percent of its energy needs. During the 1960s the United States imported about 20 percent of the oil it consumed; by the end of the 1970s it imported about 42 percent. Consequently, what happens in the Middle East, with approximately 52 percent of the world's proven oil reserves, is of major concern to all industrialized countries.

It is frequently claimed that atomic energy will soon replace oil, but the complex technological problems involved in atomic energy production, including safety and waste disposal, have been greatly underestimated. In fact, even the most optimistic forecasts show that atomic energy will provide, at best, only 8 to 10 percent of world energy requirements in the 1980s.

The strategic geographical location of the Middle East, situated at the crossroads to Asia, Africa and Europe, has made the region from ancient times the center stage of conflict among nations and empires as they vied for control over the trade routes to the East. The modern history of the region is not very different, with imperial conflict fuelled by the added attraction of the rich oil deposits discovered in the early part of this century. The construction of the Suez Canal gave Britain unchallenged control over Egypt, which the British then used as a staging ground for ousting the Ottomans from the Fertile Crescent in 1918. France collaborated with Britain against the Ottomans and together they partitioned the Fertile Crescent between them in the Sykes-Picot agreement of 1916, France laying claim to Syria and

Lebanon, and Britain taking Iraq and Transjordan, and trusteeship over Palestine. This ensured Britain's hegemony over the oilfields, set the stage for other European nations to make similar claims for oil reserves and ultimately precipitated subsequent American inroads into the area. Control over the region's oilfields was implemented through a series of concessions by which local authorities granted foreign companies long-term rights to explore, drill, extract and market local oil in return for small conditional cash payments.

This asymmetrical and unjust situation continued with only minor political interruptions until 1960, when several national governments were able to agree on the establishment of the Organization of Petroleum Exporting Countries (OPEC). This was followed by the formation of the Organization of Arab Petroleum Exporting Countries (OAPEC) in 1968. Controlling oil resources had become a central objective of Arab oil policy and was to become an integral part of the rise to international prominence of Arab political power.

Major developments in this political resurgence occurred in the early 1970s. Before October 1973 Western oil-consuming countries had assumed monopoly control over oil prices or output rates, while remaining oblivious to the inherent contradictions and injustices rampant within the oil industry. The Ramadan (October 1973) war made it possible to realize objectives which were fundamental to OPEC's establishment by revealing clearly that the 'oil crisis' was a phenomenon rooted deep within the political and economic heart of the Middle East; in other words, the oil crisis cannot be divorced from the day-to-day concerns of the Arab people, who consider the question of control over oil to be synonymous with control over their daily lives. This Arab view has emerged in response to the conceit of Western countries, which think of uninterrupted oil supplies only as a matter of life and death for their own material progress.

To repeat, oil is an inextricable part of social and political life in the Middle East. Oil touches every aspect of Arab existence, and nowhere can this be seen more clearly than in the Palestinian question in the context of the Arab-Israeli conflict. The Ramadan war was by no means the first time the Middle East had been exposed to the politics of oil. In fact, the region had been visited three times by oil 'shocks' in the previous generation – 1948, 1956 and again in 1967. During 1973 the members of OAPEC successfully demonstrated their ability to use oil as an instrument of international relations to articulate Arab interests and to achieve Arab objectives. The events that unfolded between October 1973 and January 1974 brought in their wake a restructuring of the oil industry, with the producing nations emerging as the primary determinants of oil prices. Thus the October war demonstrated, to an unprecedented degree, the extent and intensity of the links between oil and politics in the Middle East and in the world.

It is now ten years since the oil crisis of 1973. Some observers are already claiming that the oil crisis has passed from the scene. This claim is incorrect, because the oil crisis cannot be seen only in terms of an imbalance in demand for and supply of oil. As was intimated above, oil is connected by the umbilical cord of national aspirations to the Palestinian question, and as long as this fundamental problem remains unresolved, the oil crisis is alive and real. The Palestinian question ignited the October war and the war precipitated the oil crisis.

In this book, the author attempts to explain Arab oil policy and the use of oil as a political weapon to support Arab demands. His main thesis is that the oil crisis is inseparable from the Arab-Israeli conflict. The book addresses the erroneous tendency among Western Middle East specialists to separate the oil question from the Palestinian question, whether through ignorance or by design. This error is perpetuated by the claim that, even if the Palestinian question were resolved, oil prices would not decline. But the politics of oil bear directly on the Palestinian question, and oil prices are only one element in Arab political strategy. There are a number of instruments and tactics based on oil that could be marshalled in support of the Palestinian cause. The intelligent tactical use of oil in 1973 demonstrated beyond a shadow of doubt the importance of this instrument in overall Arab political strategy.

The 1973 crisis dates back to the establishment of the state of Israel in the heart of the Arab world. The history of the Arab-Israeli conflict and the oil crisis go hand in hand, and will never end until the Palestinian question itself is resolved. Arab oil power did not spring from a vacuum. It has developed as the Arab states have developed and mastered the use of oil, and Arab effectiveness in the use of oil as an instrument of diplomacy will increase with added experience and greater wealth.

It is not enough to be wealthy, however; the Arabs need to be powerful too. Power lies not in potential, but in the actual realization and implementation of one's abilities and potentialities. Power may be found in the weakest nation without being translated into regional and international influence. Power becomes influence when the capacity exists to use it effectively, and when it is used and embodied in activities capable of realizing national objectives. The relationship of oil and power is simple. Oil is energy; energy is wealth; wealth used with political acumen is the route to power.

The Arabs' possession of oil wealth dates back to the early 1930s. From the very beginning, Arab leaders realized the importance of their oil wealth as a powerful factor in world politics. They realized that oil confers political power, both in time of peace and in time of war, and that whoever controls the oilfields of the Middle East will have the power to make peace or war. However, the Arabs did not yet control their oil wealth, nor were they aware of the need for concerted action. Oil power did not become a political reality until the Arabs decided to use it during the 1967 Arab-Israeli war. Nevertheless, it was not until 1973 that the Arab oil producers became fully aware of their potential strength and were both willing and able to use it effectively.

Oil is important to the countries of the Middle East both as an instrument for economic development and as an instrument of political influence. In this book, attention is drawn to the importance of oil to the life and destiny of the Middle East, and to the international political system within which the Arab world interacts. In particular, the book examines the following subjects:

(1) The relationship between the Arab-Israeli conflict and Arab use of oil as a political instrument from the establishment of Israel in 1948 until the 1973 Ramadan war.

(2) The development of the instruments of Arab oil strategy, including cutbacks in oil production, nationalization of foreign, particularly American, oil companies, and increases in oil prices.

(3) The determining factors in the principal use of oil power, including: (a) the claim that the Arab-Israeli dispute was the primary causal mechanism, (b) the objectives which the Arabs hope to achieve from using oil power, (c) the reaction of the consumers to the Arabs' use of their oil power and (d) the impact of oil power on the attitudes of consumers in Western Europe, Japan and the United States towards the Palestinian question.

(4) The opportunities for, and limitations on, the use of oil power in the context of the present world oil market and the potential threat to oil power posed by alternative sources of energy.

In brief, the major objective of this study is to evaluate the role and significance of oil power as used by the Arab producing countries to achieve their political aims. The basic methodology and structure of the study are designed to provide a general framework within which to analyze the major forces and issues concerning oil power. In order to do so, the study deals with the international oil system by analyzing the triangular interaction among three actors: the oil companies, the oil-consuming countries and the oil-producing countries. This framework will serve as a background for understanding the Arabs' influential position in the world oil balance. Understanding the oil system is the key to understanding oil power. This study generally confines itself to the political aspects of oil power; technical, economic and geographic considerations will be treated only as they relate to the primary issue under examination.

Abdulaziz al-Sowayegh

ACKNOWLEDGEMENT

I would like to express my thanks to the Petroleum Information Committee of the Arab Gulf States for the grant they gave me to translate, update and edit the text of this book. However, I alone bear the responsibility of all the views expressed and the judgements made.

A. al-Sowayegh

PART ONE
THE OIL INDUSTRY AND THE ARAB WORLD

1 ARAB OIL: AN INTRODUCTION

Oil has not always been as important as it is today. During the nineteenth century, coal was the dominant source of energy. However, beginning with the twentieth century, the importance of oil has grown spectacularly, so that, a decade or two ago, it overtook coal as the major source of energy in the world. Within the last 50 years, while total world energy consumption has risen fourfold, world consumption of oil has risen by a factor of 16. Currently, gas and oil account for almost 70 percent of world energy consumption. The energy transition from coal to oil was partly a response to technological developments, but even more significant was the steady decline in the real price of oil.

From 1870 onwards, world oil production rose from an initial production rate of 10 million tons in 1872 to 100 million tons in 1890, and then almost doubled every ten years, from 1,000 million tons in 1960, to 2,000 million tons in 1970, and to over 3,500 million tons in the early 1980s.

Although oil reserves are known to be much less abundant than coal — there exist about 150 billion tons of oil compared to 700 billion tons of coal — private and public energy policies in the West have generally favored the exploitation of oil at relatively low cost, especially from Arab oil producers. The net result has been rapid economic development in the West supported by cheap energy, but at the expense of developing alternative energy sources. It has also meant that high growth rates were achieved through the creation of an energy-intensive system of production, distribution and transportation.

Arab oil occupies a critical and central position in the world energy situation. In 1980 Arab proven reserves exceeded 333 billion barrels, or over 52 percent of the world total. Arab production of oil in the same year reached a peak of 19.5 million barrels per day (b/d) or 36.3 percent of total world production. As a percentage of total OPEC production, the Arabs' share exceeded 71 percent. The largest Arab shares, however, are those in the world trade in oil: Arab oil exports ranged between 83 and 87 percent throughout the 1970s. The combination of large pools of reserves, substantial production rates, and world dependence on their exports of oil catapulted the Arabs into the center of world events, as nations competed either to control Arab oil-wells directly or to bring the Arab nations into their spheres of influence.

3

Lord Curzon's famous statement 'The Allies floated to victory on a sea of oil' sums up well the strategic value of oil in the world balance of power. It comes as no surprise that competing world powers have had an overwhelming interest in the Middle East. Since the discovery of oil in the area in the early 1920s, scarcely any world power has directed a policy towards the Middle East that has not incorporated an oil concern or interest.

Imported oil accounted for over two thirds of the domestic availability of oil in the Organization for Economic Co-operation and Development (OECD) countries in 1980, and Arab oil contributed about 71 percent of European and about 47 percent of United States oil imports in the same year. It follows that developments in the Arab oil industry, and developments outside the oil sector but relevant to oil flow, prices and distribution, are of vital concern to the West and to the world at large.[1] Moreover, the Arab region, because of its oil revenues, is also a major market for imported industrial products. Furthermore, Arab financial surpluses are the major source of funding the balance of payments deficits of many countries in addition to being a major source of investment finance in the Third World.

Arab imports in 1980 exceeded $122.7 billion. This figure excludes the lucrative invisible trade items but nonetheless it accounts for about 10 percent of world imports in this year. The Arab world imported almost 44 percent of its requirements from the EEC countries and more than 10 percent from each of the United States and Japan. At the same time, the Arabs imported $12 billion from developing countries, almost 70 percent of the total exports of these Third World countries in 1980.[2]

The growing volume and value of oil exports necessarily implied an accumulation of financial surpluses, coming as they did at a time when the import absorptive capacity of Arab economies was limited. These funds are now believed to total $300 billion, and are held primarily in Western financial institutions. This formidable convergence of burgeoning oil revenues, a growing market for imported industrial products and command over huge financial resources has raised the world's economic and political stakes in the Arab nation. Oil is the cornerstone, but each element will play a fundamental role in Arab strategy.

Notes

1. Data on energy imports are from OECD, *Energy Statistics* (OECD, Paris, 1982).

2. The trade data are all from United Nations, International Monetary Fund, *Director of Trade (1981)*.

2 MAJOR ACTORS IN THE OIL GAME

The Oil Companies

The oil industry operates through a complex network of relationships spreading throughout the world. It is the world's leading industry in size; it is probably the only international industry that concerns every country in the world. As a result of the geographical separation of regions of major production from regions of high consumption, oil is the major traded commodity and dominates international trade in tonnage and value.[1]

Until very recently, however, ownership and control of the oil industry were in the hands of a very small group of companies which dominated the whole industry.[2] These companies can be categorized into two broad groups: 'Majors' and 'Independents'.

(1) The seven leading international oil conglomerates are often referred to as the 'Seven Sisters', the 'International Majors', or simply the 'Majors'. Of these seven, five are American, one is British, and one is Anglo-Dutch. The members are: (a) Standard Oil of New Jersey (Exxon),[3] the largest company and often seen as the prototype of the international trust; (b) the Royal Dutch-Shell Group (Shell); (c) British Petroleum Company (BP);[4] (d) Gulf Oil Corporation (Gulf); (e) Texas Oil Company (Texaco); (f) Standard Oil Company of California (Socal or Chevron); and (g) Mobil Oil Company (Mobil). An eighth company, Compagnie Française des Pétroles (CFP), has since gained in international importance and is now often added to the Majors, although it is smaller than the other members.[5] Table 2.1 gives the sales and assets of the Majors, whereas Table 2.2 presents their shares of oil production in the Gulf in 1971 and Figure 2.1 shows the pattern of ownership that existed in the Gulf before 1974.

(2) The international 'Independents' or 'Newcomers' consist of about 20 to 30 relatively small private and state-owned companies which entered the oil arena during the 1950s and 1960s. Nonetheless, their sphere of operation is slowly widening. Such firms as Standard Oil of Indiana, Phillips Petroleum, Continental Oil, Getty, and Ente Nazionale Idrocarburi (ENI) are included in this group.

6

Table 2.1: Sales, Assets, Employees and Net Income of Major Oil
Companies Outside the United States, 1972 and 1982[a]

Company	Sales[b]		Assets[b]		Employees		Net income[c]	
	1971	1981	1971	1981	1971	1981	1971	1981
Exxon	18.7	39.7	20.3	20.1	143,000	41,570	1,532	1,099
Shell Group	12.7	82.3	19.7	65.3	185,000	166,000	704	3,642
Mobil Oil Group	8.2	9.5	8.6	4.6	75,300	8,792	574	153
Texaco Oil	7.5	8.0	10.9	4.3	75,200	9,692	889	370
Gulf Oil Group	5.9	4.7	9.5	3.8	57,200	12,306	197	311
British Petroleum	5.2	52.2	7.9	47.9	70,600	153,250	178	2,063
Socal or Chevron	5.1	3.2	7.5	0.7	42.500	1,639	547	26

Notes: a. Entries in the table exclude US operations.
b. In billion US dollars.
C. In million US dollars.
Source: *The Fortune Directory of the Largest Industrial Corporations Outside the
United States (Fortune, 1972 and 1982).*

The history of the international oil industry runs parallel to the
history of the international oil companies. For many years, the major
international and multinational companies dominated the world oil
industry. They financed the risks of exploration, determined the rate
of production, set the price of products, delivered them all over the
world and even sold them at the pump. In short, they controlled every
aspect of the oil industry, particularly in the Middle East.

No one has explained the rise to power of the Majors more effectively than Enrico Mattei. In a speech given in 1959, he said:

That the oil industry has become increasingly concentrated in the
hands of a small group of vertically integrated organisations which
are banded together in various ways, and that for a long time
satisfaction of the oil needs of the whole world has almost entirely
depended on these, and still does to a large extent today, is due to
the conditions in which the oil industry began and grew up.

Undoubtedly the system met military needs and those of a world
economy in the developing stage. But it is equally undeniable that
consumers of petroleum products have paid dearly for the services
rendered.

It is worth noting that for the very small number of large international companies — most of them also engaged in the oil industry
in the United States — dominating world crude oil production, there
was every interest in exploiting the resources available in all areas,

Table 2.2: Oil Output of Major Oil Companies in the Gulf, 1971 (million metric tons)

Gulf OPEC countries	Esso	Shell	BP	Gulf	Texaco	Socal	Mobil	CFP	Other	Total
Iran	14.5	29.1	83.8	14.5	14.5	14.5	14.5	12.5	26.3	227.0
Saudi Arabia	66.6	–	–	–	66.6	66.6	22.2	–	–	222.0
Kuwait	–	–	72.5	72.5	–	–	–	–	–	145.0
Iraq	9.8	19.7	19.7	–	–	–	9.8	19.7	4.3	83.0
Abu Dhabi	3.3	6.5	17.8	–	–	–	3.3	12.2	1.4	44.6
Neutral zone	–	–	–	–	–	–	–	–	26.4	26.4
Qatar	1.1	12.5	2.4	–	–	–	1.1	2.4	0.5	20.0
Total	95.2	67.8	196.2	87.0	81.1	81.1	50.8	46.8	58.9	764.9

Source: OPEC, *Annual Statistics Bulletin, 1972.*

Figure 2.1: The Network of Ownership of Major Oil Companies in the Gulf prior to 1974

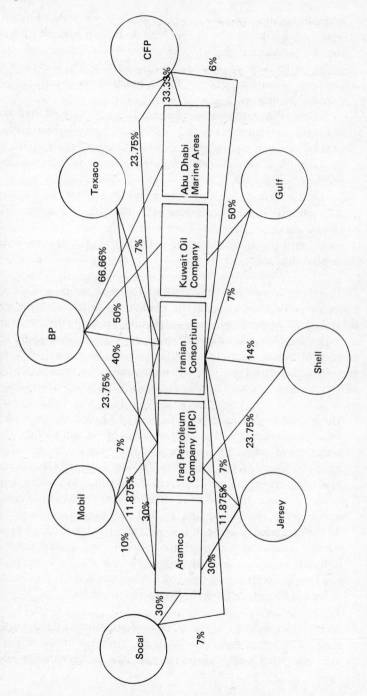

Source: Edith T. Penrose, *The Large International Firm in Developing Countries* (M.I.T. Press, Cambridge, Mass., 1968), p. 151.

without boosting output in any of them in particular. For some thirty years now these companies have, in fact, functioned as regulators of production, sharing coverage of the world demand determined by the level of prices and by the fiscal policy of each of the consumer countries between the different areas and the various producer companies.

Clearly, this state of affairs has not arisen from a concerted plan, but a set of circumstances and a network of tacit agreements have had the outcome at any given moment of eliminating competition between the big companies and potential challenge to them from smaller organisations.

But the system has also been based on interests of the major international oil groups coinciding with the national interests of the United States of America, namely that the selling price of crude should be high enough to make extraction profitable even in continental America.[6]

The Majors controlled, separately or together, the three main processes of production, refining and marketing. As late as 1968, they controlled 77.9 percent of world production, excluding the communist countries and North America, 60.9 percent of world refining, and 55.6 percent of world marketing facilities. They also controlled a vast majority of known oil reserves. Just a few years earlier these percentages were even higher.

These companies have built refineries which dot the earth from Japan and Australia to the major countries of Western Europe. Their tanker fleets roam the seven seas, dispatching crude and refined products, from the [Persian] Gulf, Indonesia, Venezuela and other producing centers to markets everywhere, while their prices march in majestic harmony, and their division of the world's production rests in reasonable equilibrium. In marketing they permit themselves the extravagance of competition, so that a good part of the economics of massive organization are frittered away in duplication of facilities, as in the domestic market in the US. Their animosity, however, is directed not against each other, but against consumers of all lands who exist to fatten their imperial treasuries.[7]

In recent years, because of the oil-producing countries' determination to have more say in their oil operations, the share of the Majors in various aspects of world oil transactions has declined considerably.

Table 2.2 illustrates this trend.

Although these changes occurred abruptly, they were occasioned by several developments in the framework in which the international oil companies had operated during the previous few decades. As early as 1950, many countries had begun to pay more attention to the activities of multinational corporations. For the oil-producing countries, attention was naturally drawn to the operations of the Majors. By 1960 OPEC had been established and it had become clear that 'the conditions under which bilateral monopoly bargaining [was] being carried out between companies and governments [were] changing in favor of the latter', and that 'an increasingly large share of the economic rent derived from the low production costs in the Middle East and from the price structure [would] be absorbed by the governments'. By the mid-1960s it was clear that 'the twin pillars which held the conduit — Anglo-American military and political power in the Middle East and North Africa and control of production by the American and British companies — [were] being undermined'.

Many factors contributed to the decline in the major oil companies' control over the international oil industry. First, the 1960s witnessed declining profit rates in the international oil industry, partly because increased competition, triggered by Newcomers, drove down the price of crude oil. The rapid growth of European oil enterprises such as CFP of France and ENI, the Italian state-owned company, proved effective in neutralizing the absolute power of the Majors. With the success of these Newcomers in obtaining major concessions in the Middle East and other areas, it became obvious that the monopoly power of the Majors was deteriorating. Even more important were the activities of the American oil companies which used to operate only domestically but which, beginning in the early 1920s, sought both oil reserves and markets abroad. Their success in obtaining foreign reserves and foreign markets for their oil supplies, particularly in Western Europe, was partly at the expense of the Majors.

Second, the growing incidence of environmental problems and heightened public awareness of environmental pollution, particularly in the United States but also in Japan and Western Europe, added considerable uncertainty about future alternative energy sources to oil. Thus the Majors were hesitant to expand their operations in the oil-producing countries until they could be certain of the impact of environmental priorities on energy choices by policy-makers in the industrialized countries.

Third, the wave of nationalizations, especially of oil operations, by some oil-producing countries in the Middle East and North Africa made these countries increasingly less dependable as secure sources of oil for the Majors.

Finally, the weakening of the West, which had formerly dominated the Middle East's large oil resources, was a major explanatory factor in the diminishing influence of the Majors.

These circumstances endangered the major international oil companies by threatening to deprive them of their monopolistic control over low-cost crude oil supplies, and by threatening to make the control less profitable if other energy sources gained a competitive advantage over oil. However, in response to this threat, the Majors adopted strategies which sought to restore their monopolistic position and profitability by gaining control over *all* energy sources and *all* downstream operations. To implement this strategy a number of tactics were employed.

The first and most vital step was to gain control over oil-related areas of operation. In the 1950s and early 1960s, the Majors moved significantly into the petrochemical industry. By 1962, for example, the Majors owned or operated over one third of the petrochemical plants in the United States.[8] This process accelerated during the late 1960s and early 1970s.

Furthermore, the Majors broadened their operations, becoming 'energy' companies, as they bought controlling interests in competing energy resource industries, such as coal, uranium, and even exotic sources like geothermal energy. During their earlier domination of the oil business, the giant oil companies had learned that their greatest advantage was as monopolistic suppliers of raw materials, because control of raw materials is an effective bar to new entry into an industry, thereby ensuring very high rates of return on investment.

The major oil companies have not yet abandoned their historical dominance of the production and downstream phases of the oil industry. However, in recent years, their dominance has declined significantly, as the introduction of new elements and participants in both the economic and the political sphere has reduced their share of the spoils.

In the 1960s, the introduction of the principle of 'participation' shifted control of oil exploitation and trade away from the leading oil companies to the governments of the producing countries. However, these oil companies have remained chief participants in the oil industry, first, because continuing to do business with the Majors is profitable for the oil-producing countries and, secondly, because the consuming countries continue to demand that the oil companies participate in the oil trade. The international oil companies, in effect, continue to act as intermediaries between the oil-producing and the oil-consuming countries. However, such a reduced role has curtailed the freedom of action of these companies and, combined with government regulations, augurs their elimination as a 'buffer between the exporting and consuming countries'. This process has already begun to unfold. As Sheikh Ahmed Zaki Yamani conceives it:

[The oil-producing] governments now retain entirely the decision-making powers of price determination, production levels in their countries, future expansion of oil facilities and, to a great extent, the destination of oil exports. Consequently, the role so successfully played by the oil companies as bridge and buffer between the exporting and consuming countries has diminished, giving rise to the need for filling the vacuum thus created by a new two-party relationship between the two groups of countries. *The role being played by the oil companies now is properly that of a purchaser, refiner, and provider of technology.*[9]

The Oil-Consuming Countries

The second major group in the international oil triangle consists of the oil-consuming countries. Within this group we can differentiate between two forces: (a) the home governments of the international oil companies; and (b) the countries comprising the oil-importing Third World.[10]

The Home Governments of the International Oil Companies

The home governments of the international oil companies have a special status in the international oil industry. All the major oil companies are owned by residents of major Western powers. The United States is by far the most important; five of the seven Majors and the majority of the emerging Newcomers are domiciled there. The United Kingdom is the second most important Western oil power, having a government-owned majority share in BP and a 40 percent share in Royal Dutch–Shell. France owns two oil companies, one the partially government-owned CFP (which may be considered the eighth Major) and the second the state-owned ERAP (Entreprise de Recherches et d'Activités Pétrolières). Italy has ENI, and Japan has Arabian Oil Company. Other West European nations own some important Newcomers, but none of them owns a major producer.

The history of the international oil industry begins with the history of American production. From 1859 when Edwin L. Drake drilled his first well, to 1910 when the first American wells were drilled abroad, the oil industry developed almost exclusively in the United States. Rockefeller, through Standard Oil, was the core of the American oil industry, much in the same way that the major oil companies are the core of the contemporary international oil industry. As Harvey O'Connor has stated:

> For a half century the history of oil was also the personal history of John D. Rockefeller, who tamed an anarchic industry and brought it under the direct control of Standard Oil. The oft-told tale ran the spectrum of the devices of monopoly. Competitors were bought out or ruined, legislators and public officials were also bought out (and many ruined), laws were flouted with impunity or by stealthy indiscretion.[11]

Until the early twentieth century, the United States had little interest in and no control over foreign oil production. As late as 1914, American firms had acquired producing properties only in Mexico and Romania, and foreign production by United States nationals accounted for only 15 percent of the total crude oil output outside the United States.

The First World War brought to light the importance of oil as a strategic material and underscored its necessity to the world economy. The lessons taught by the war made the great powers more oil-minded than ever before and heightened their competition over oil, placing it in a new international perspective. Moreover, these nations also came to realize that oil's importance was not limited only to times of war; rather, oil was equally important in times of peace, although the emphasis in peace-time was more economic than military.

The growing importance of oil made it imperative to ensure its availability. Increasing demand for oil made its production and distribution an attractive commercial proposition. The struggle for oil was never as hotly contested as after the First World War.

Spurred on by local panic over a domestic shortage of oil supplies, American oil companies with government encouragement started to explore potential oil-bearing lands abroad. The 1920s witnessed a great expansion of American foreign drilling concessions, as the emphasis shifted from 'home' reserves to the acquisition of foreign oil resources. However, the foreign arena was not without its complications. In particular, American companies were confronted by a strong and determined competitor, the United Kingdom, which had learned from the war that secure oil supplies were vital to its survival.

Before the outbreak of the First World War, British leaders were alarmed over Britain's oil predicament; all British supplies had to come from far-away areas. In the event of war, Great Britain would have to establish safe transportation routes for oil from great distances. To overcome these difficulties, two different policies were suggested. The first policy, advocated by Lord Fisher (sometimes called England's 'oil maniac'), was based on playing off Royal Dutch-Shell, the Anglo-Persian group, the Mexican Eagle, the Burma Oil Company and the American companies against one another by purchasing oil at opportune times and placing it in storage at strategic bases for the use of the Admiralty.

However, another oil policy, advocated by Winston Churchill, then First Lord of the Admiralty, provided for the Admiralty's purchase of a controlling majority share in British oil companies and, through such companies, the purchase of oil rights and concessions in foreign countries for the purpose of ensuring oil supplies for Britain and the Admiralty. Churchill put forward this policy, which later became the basis of Britain's oil strategy, before the House of Commons on 17 July 1913:

It is a two-fold policy. There is an ultimate policy and there is an interim policy. Our ultimate policy is that the *Admiralty should become the independent owner and producer* of its own supplies of liquid fuel, first, by building up an oil reserve in this country sufficient to make us safe in war and able to override price fluctuations in peace; secondly, by acquiring the power to deal in crude oils as they come cheaply into the market . . . This second aspect of our ultimate policy involves the Admiralty being able to refine, retort, or distill crude oil of various kinds, until it reaches the qualities required for naval use.

The third aspect of the ultimate policy is that *we must become the owners, or at any rate, the controllers at the source*, of at least a proportion of the supply of natural oil which we require. On all these lines we are advancing rapidly . . . The interim policy consists in making at once a series of forward contracts . . . and thirdly, to draw our oil supply, so far as possible, from sources under British control or British influence, and *along those sea or ocean routes which the Navy can most surely protect.*[12]

(Emphasis in the original)

The British government moved rapidly to carry out this policy. The policy theater was that of the Middle East. Britain had long had a pervasive interest in the Middle East as the strategic gateway to India and the Orient, and it was Britain which discovered the area's oil potential. In Mosul, the British had a one quarter interest with the Germans and Turks in the Turkish Petroleum Company (TPC). Within the span of a few months, the British Cabinet had increased British ownership in the company to three quarters, leaving the Germans with only a one quarter interest and shutting out completely the Turks who controlled the territory, the Iraqis who owned the oil lands, and the Americans who held the concession.

That was only the beginning. In 1913 a mission despatched to Persia by Winston Churchill recommended that the British government provide financial support for the expansion of the Anglo-Persian Oil Company (APOC) and have 'a voice in the direction of the Company's general policy'. On 30 May 1914 Churchill's recommendation was carried out and he signed a contract with APOC under which the British government bought, for $11 million, a controlling interest in the Anglo-Persian Company. Two directors were to be appointed to APOC's board; they represented the Admiralty and had veto rights over all strategic issues. With this contract went 48 years of monopoly over most of the Persian Empire, possessing the richest known oilfields of the Eastern hemisphere at that time.

After the armistice agreement, the British government set out to put Churchill's plan into action with the aim of achieving British control over Middle Eastern and Latin American oil resources. However, this time 'the struggle for oil [was] no longer a rivalry between great trusts; it [was] a struggle between the nations'. Thus, after the First World War, the struggle for oil entered a new era of confrontation.

To achieve its goals, Britain collaborated with France and the Netherlands with the immediate objective of undermining both the American oil companies' position on foreign soil and the American government's prestige abroad. The British Foreign Office strengthened its diplomatic ties to defend and extend its claim to concessions in the Middle East and elsewhere. British companies were encouraged to become more aggressive in seeking and obtaining oil concessions in foreign countries.

The main challenge to Standard Oil was launched by Royal Dutch-

Shell.[13] In addition to Shell's activities, two organizations were chosen to pursue the drive to obtain new foreign concessions. These were the D'Arcy Exploration Company, an Anglo-Persian subsidiary which the British government owned directly, and British-Controlled Oilfields, Ltd., a company incorporated under Canadian law which was organized specifically to close Central and South American countries to the United States. Meanwhile, the United States government was struggling to gain entrance for its nationals to the British-dominated Middle East.

On another front, Britain and France signed the San Remo Agreement on 24 April 1920. The object of this agreement was to exclude the United States from all foreign oilfields and markets under British and French influence and from the mandated territories taken from the Central powers and their allies. The Americans were outraged. In May 1920 President Wilson submitted to the Senate a report from the Secretary of State describing discrimination against American citizens who wished to explore for oil abroad. In July of the same year, the US State Department observed: 'It is not clear to the Government of the United States how such an agreement can be consistent with the principles of equality of treatment understood and accepted during the peace negotiations at Paris.'[14] In 1921 the Secretary of the Interior informed Senator Henry Cabot Lodge that Great Britain had prevented Americans from operating in any British-controlled oilfield and had placed heavy financial burdens upon them.

The American oil companies, supported by American public opinion and Congress, started a furious campaign to persuade Washington to interfere on their behalf against 'this iniquitous carve-up'. Under tremendous pressure, the State Department struck hard and fast. Diplomatic notes shot back and forth between Secretary of State Robert Lansing and his British counterpart, Lord Curzon. America insisted on a share in the fruits of victory. In the words of State Department officials: 'This Government has contributed to the common victory, and has a right, therefore, to insist that American nationals shall not be excluded from a reasonable share in developing the resources of territories under mandate.'[15] Though accepting the principle, Britain maintained that it did not apply to rights which had been granted before the war.

The 'open-door' policy was a high priority for American diplomats; it was based upon equal rights, equal opportunity, square dealing and good sportsmanship. The most significant action taken by the United States government to implement the open-door concept was finally resolved only when France cancelled the San Remo Agreement in 1928.

Thus in 1928, after many years of struggle, American companies finally broke through the stubborn resistance of the British, Dutch and French, and secured their first substantial interest in the potential oil resources of the Middle East. The British, French and Americans agreed on their share-holdings in the Iraq Petroleum Company (IPC).[16] Thus the open-door policy was finally initiated. As Leonard Mosley put it, by the 1928 agreement, 'the British had finally opened the door to the Middle East oil fields and let the Americans put one foot inside. From now on the struggle would be to keep them from bringing in the other.'[17]

Thus, the State Department played an important part in bringing the American oil business into the Middle East and paved the way for American influence in the Middle East during the post-1945 era. Jack Anderson's brilliant study of many State Department confidential papers concluded in 1967:

> *[T]he State Department has often taken its policies right out of the executive suites of the oil companies.* When Big Oil can't get what it wants in foreign countries, the State Department tries to get it for them. In many countries, the American Embassies function virtually as branch offices for the oil combine . . .
>
> The State Department can be found almost always on the side of the 'Seven Sisters', as oil giants are known inside the oil industry.[18]

Some viewed these successive intercessions on behalf of oil concerns as fulfilling the 'moral' purpose and political needs of the natural overseas extension of nineteenth-century manifest destiny. In reality, it was a naked example of *powerpolitik* in action. More significantly, these moves proved that the United States and its corporate citizens were developing as a world power, with continuing economic and political interests on every continent.

The Oil-Importing Third World

The second force within the oil-importing countries is composed of the Third World non-oil-producing members. These countries are the most sensitive to changes in the international oil industry, because they are less able than most other countries in the world to absorb the costs of adjustment to higher energy prices and endure shortages in available energy supplies. Besides, as the data in Table 2.3 show, energy demand in these countries is expected to rise substantially in the 1980s. The World Bank puts the anticipated annual rate of growth at 6.31 percent throughout

Table 2.3: Commercial Primary Energy Consumption by Country Group, 1977–90 (million barrels of oil equivalent per day)

Country group	1977	1980	1985	1990	Average annual rate of growth 1977–1980	1980–1990
Industrialized countries	70.6	70.1	80.7	92.5	−0.24	2.74
Centrally planned economies	40.3	42.5	51.8	64.3	1.79	4.23
Capital-surplus oil exporters	1.6	1.9	2.7	3.9	5.60	9.32
Developing countries	17.1	18.7	25.0	34.4	1.66	6.25
Net oil exporters	3.9	4.6	6.0	8.3	5.66	6.08
Net oil importers	13.2	14.1	19.0	26.0	2.22	6.31
Bunkers and others	4.6	4.6	5.8	6.5	—	—
Total	134.2	137.8	166.0	201.5	0.89	3.97

Sources: UN, *World Energy Supplies, 1973-78* (Series J, no. 22); World Bank estimates; and High-case projections of *World Development Report, 1980*.

the 1980s. In fact, a number of indicators already show that the energy requirements of developing countries are on the increase. One such indicator is the energy coefficient, which is the ratio of the percentage change in energy demand to the percentage change in GNP. The developing countries' energy coefficient, which used to average 0.76 in the period 1948 to 1973, rose to 1.25 between 1975 and 1980, and is expected to rise further in the eighties. This contrasts with declining energy coefficients for the developed countries. The efforts of the developing countries to industrialize are considered the main factors explaining the rise in the energy coefficients. However, the rise in the cost of energy should not be neglected as another important explanatory factor.

Table 2.4 presents a detailed classification of most of the developing countries in terms of the ratio of oil imports to their corresponding total energy demand. Invariably, the poorest developing countries are also the most dependent on energy imports. Central to these common problems are several interrelated vicious circles of under-development in the energy and petroleum resources areas, which limit these countries' potential for economic development. As Michael Tanzer summed them up, the major energy-related constraints facing these countries are:

First, industrialization and modernization generally require even more rapid growth in energy consumption than in the total output of the economy. Second, many of the countries have a shortage of indigenous energy resources, due either to nature, lack of knowledge as to the resource base, or failure to exploit proven resources. Third, most of the countries are short on foreign exchange which could be used for importing additional needed energy, particularly petroleum. Fourth, many of the countries rely to a lesser or greater degree on foreign aid to help fill their foreign exchange gap. Fifth, and finally, their internal petroleum sector, whether well or poorly developed, is generally dominated by the international oil companies whose home governments, particularly the United States, are the primary sources of this foreign aid.[19]

Figure 2.2 shows the net oil imports of non-OPEC Third World countries for 1977. India is typical of these developing countries and its recent economic experience exemplifies the adverse impact of changes in the international oil industry. The constraints placed on India's developmental strategy by OPEC's success in raising world oil prices illustrate the limited freedom of action open to many Third

Table 2.4: Classification of Developing Countries by Intensity of Energy Demand

Net imports of oil as a percentage of energy demand	Oil exporting developing countries		Oil importing developing countries			
	OPEC members	Other than OPEC members	0–25%	26–50%	51–75%	76–100%
	Algeria Gabon Iran Iraq Kuwait Libya Saudi Arabia UAE Venezuela	Bahrain Bolivia Malaysia Mexico Oman Peru Syria Trinidad & Tobago Tunisia	[Argentina] [Columbia] Democratic Korea [Romania] South Africa	[Chile] Mongolia Yugoslavia	[Albania] [Brazil] Republic of Korea Lebanon Turkey	Bahamas Barbados Costa Rica [Cuba] Cyprus Dominican Republic Fiji [Guatemala] Guyana [Ivory Coast] Jamaica Jordan Malta	Mauritius Nicaragua Panama Papua New Guinea Paraguay Portugal Surinam Uruguay
The countries that face actual and expected problems concerning wood fuel	Equador Indonesia Nigeria	Angola Burma China Popular Congo Egypt Zaire	[India] Vietnam Zimbabwe	[Bangladesh] Botswana Mozambique [Pakistan]	Afghanistan Burundi [Ghana] Malawi Ruanda	Benin Bhutan Cameroon Cape Verde Central African Republic Chad Comoro El Salvador Equatorial Guinea Ethiopia Grenada Guinea Bissau Haiti Honduras Cambodia Kenya Lao People's Democratic Republic Liberia Madascar Maldives Mali	Mauritania [Morocco] Nepal Niger Lesotho Senegal Solomon Islands Sierra Leone Somalia Sri Lanka Sudan Swaziland Tanzania [Thailand] Togo Uganda Upper Volta North Yemen South Yemen Djibuti

Source: UN *World Energy Supplies, 1973-78* (Series J, no. 22); and World Bank, *World Development Report, 1980.*

Figure 2.2: Net Oil Imports of Non-OPEC Third World Developing Countries, 1977

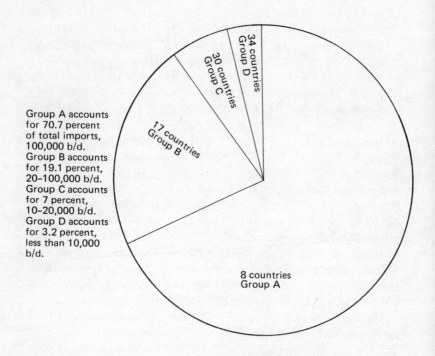

Group A accounts for 70.7 percent of total imports, 100,000 b/d.
Group B accounts for 19.1 percent, 20–100,000 b/d.
Group C accounts for 7 percent, 10–20,000 b/d.
Group D accounts for 3.2 percent, less than 10,000 b/d.

34 countries Group D
30 countries Group C
17 countries Group B
8 countries Group A

Source: OAPEC, *Studies in Arab Oil Industry* (Kuwait, 1981), p. 479.

World countries. It should be remembered, however, that India, unlike many developing countries, is relatively rich in mineral resources, although much of this resource endowment remains to be exploited. The growing imbalance between India's fuel endowment and fuel demands was met by energy imports, which further strained its balance of payments position. In 1974 India's total oil imports were estimated at 17 million tons. Of these, 9.55 million tons came from Iran, 2.80 million tons from Iraq and 3.85 million tons from Saudi Arabia. According to the calculations of the *Petroleum Economist*, India had to pay $1,241 million for its oil imports in 1974, which was equivalent to 40 percent of its potential receipts from exports and twice the amount of its foreign currency reserves. As S. Manoharan, of the Indian Institute of

Foreign Trade, put it:

> There are reasonable doubts about the [Indian] government's ability to find the required foreign exchange resources for importing the sizeable quantity of 17 million tons of oil which, at an average price of $10 per barrel, will cost nearly Rs. 9,000 million excluding freight charges. There would be an additional bill in respect of petroleum products like kerosene which is in deficit and which has to be imported. The total cost may, therefore, be easily Rs. 12,000 million against about Rs. 2,040 million in 1972-74.[20]

It is clear from the foregoing discussion that India, and for that matter all developing and developed countries, must take initiatives to curtail waste in energy consumption. Energy supplies are finite and non-renewable and therefore must be used sparingly and wisely. The fact that developing countries will need far more energy supplies than has hitherto been the case is all the more reason for these countries to plan their energy uses judiciously in order to maximize the contribution of energy to their industrialization efforts. Their development plans will, of necessity, depend to a large extent on their ability to establish priorities in their energy consumption, and their ability to adjust to and comply with the new reality of limited, costly energy supplies.

On the other hand, the oil-producing and -exporting countries, particularly the Arabian Gulf countries, have contributed substantial financial aid to their brothers in the Third World. This aid is unique as it represents a transfer of resources from some developing countries to other developing countries. The main donors among oil-exporting countries are admittedly endowed with large financial assets; yet this wealth is derived from a depletable natural resource, not from industrial power. Aid is given from oil revenues which represent, to a large extent, the monetary realization of a depletable capital asset. The benefits derived by the recipients of this aid are greater than is apparent from the nominal dollar value because OPEC grants and loans are rarely, if ever, tied to source.

By the early 1970s aid from OPEC countries was already substantial, amounting to $688.9 million in 1973, but their disbursements increased to a staggering rate thereafter. In 1980, for example, the Arab members of OPEC alone disbursed $6.8 billion, of which over 87.9 percent was in the form of grants (see Table 2.5). As a percentage of the combined GNP of the Arab donor countries, Arab net disbursements accounted for 2.34 percent. The remarkable generosity of the Arabs is revealed by

Table 2.5: ODA Performance of Individual Arab Countries

Donor country		1973	1974	1975	Net Disbursements 1976	1977	1978	1979	1980
Saudi	A	334.9	1,622.1	2,466.7	2,817.3	2,709.7	1,746.7	2,298.0	3,040.0
Arabia	B	4.4	7.2	6.7	6.7	4.9	2.8	3.0	2.6
	C							62.1	86.7
	D							20.2	3.1
Kuwait	A	555.7	1,187.1	1,712.2	1,875.7	1,864.4	1,150.8	1,055.0	1,188.0
	B	9.2	10.9	14.2	13.3	13.0	6.1	4.1	3.9
	C							74.3	87.8
	D							8.4	3.4
UAE	A	288.6	749.4	1,206.6	1,144.5	1,395.2	761.4	1,115.0	1,062.0
	B	16.0	11.1	16.3	11.9	12.1	6.6	5.9	4.0
	C							97.2	86.1
	D							1.2	1.4
Iraq	A	11.1	440.2	254.4	254.7	135.2	211.2	847.0	829.0
	B	0.2	4.2	1.9	1.6	0.7	1.0	2.5	2.1
	C							81.8	87.7
	D							0.8	1.6
Qatar	A	93.7	217.9	366.7	240.3	265.9	133.6	277.0	319.0
	B	15.6	10.9	16.9	9.8	10.7	4.6	5.9	4.8
	C							97.0	99.5
	D							0.6	1.6
Libya	A	403.8	263.2	362.8	363.2	287.0	548.9	105.0	281.0
	B	6.3	2.2	3.2	2.5	1.6	3.0	0.5	0.9
	C							100.0	97.9
	D							8.6	4.8
Algeria	A	29.8	51.4	42.2	66.6	73.1	55.2	272.0	83.0
	B	0.3	0.4	0.3	0.4	0.4	0.2	0.9	0.2
	C							77.4	86.0
	D							1.5	5.2
Total	A	1,717.6	4,531.3	6,411.6	6,762.3	6,730.7	4,607.8	5,969.0	6,802.0
	B	4.2	5.4	6.3	5.7	4.8	2.8	2.9	2.1
	C							76.7	87.9
	D							9.9	2.7

A = ODA net disbursements in million US dollars.
B = ODA as percentage of GNP.
C = Overall grant element of ODA commitments (as percentage).
D = Share of worldwide multilateral organizations in total ODA (as percentage).
Source: OECD, *Development Co-operation*, various issues.

comparison of US aid, for example, which fell below 0.27 percent of
American GNP in 1980. For some individual Arab countries, aid
contributions were much higher: in 1975 aid as a percentage of GNP
was 15.62 percent in Qatar, 11.68 percent in the United Arab Emirates
(UAE), 8.11 percent in Kuwait and 5.62 percent in Saudi Arabia (see
Tables 2.5 and 2.6). Saudi Arabia's official aid was almost as high in

Table 2.6: Concessional Assistance by Arab Countries, Net Disbursements (million US dollars)

Country	1970	1971	1972	1973	1974	1975	1976	1977	1978	1979	1980
Algeria	–	–	–	25	47	41	54	47	44	272	83
Iraq	–	–	–	11	423	218	232	61	172	847	829
Kuwait	148	108	–	345	632	976	621	1,517	1,270	1,055	1,188
Libya	63	53	64	214	147	261	94	115	160	105	281
Qatar	–	–	–	94	185	339	195	197	100	277	319
Saudi Arabia	155	160	204	305	1,029	1,997	2,415	2,410	1,719	2,298	3,040
UAE	–	50	74	289	511	1,046	1,059	1,238	717	1,115	1,062
Total	366	371	494	1,283	2,974	4,878	4,670	5,585	4,187	5,988	6,802

Concessional Assistance by Arab Countries, Net Disbursements as Percentage of GNP

Country	1970	1971	1972	1973	1974	1975	1976	1977	1978	1979	1980
Algeria	n.a.	n.a.	n.a.	0.28	0.37	0.28	0.33	0.24	0.18	0.87	0.21
Iraq	–	–	–	0.21	3.98	1.65	1.45	0.33	0.76	2.53	2.12
Kuwait	5.81	3.50	4.11	5.72	5.81	8.11	4.56	10.02	7.37	4.09	3.88
Libya	1.86	1.33	1.38	3.32	1.26	2.31	0.63	0.65	0.93	0.45	0.92
Qatar	n.a.	n.a.	n.a.	15.67	9.25	15.62	7.95	7.91	3.57	5.89	4.80
Saudi Arabia	5.02	3.76	3.66	4.04	4.45	5.62	5.15	4.10	2.66	3.01	2.60
UAE	n.a.	3.85	4.63	12.69	7.05	11.68	9.21	8.49	6.05	6.87	8.96
Total	4.04a	2.94a	3.18a	3.46	3.80	4.99	3.84	3.80	2.56	2.79	2.34

Note: a. Average of the aid-giving countries.
Source: OECD, *Development Co-operation*, 1981 Review.

absolute dollars as that of Japan or West Germany in 1980. When aid contributions by Saudi Arabia, Kuwait and the UAE are combined, they rank second only to the US in absolute dollars. Algeria even gave aid to several developing countries with *per capita* incomes higher than its own.

To these substantial aid flows can be added almost $10 billion in remittances paid by foreign workers in the Arabian Gulf to their families in other Arab countries, India, Pakistan, Bangladesh, Sri Lanka, and so forth. Another $12.3 billion in 1980 represented purchases by Arab oil-producing countries from the oil-importing developing countries. The total sum of almost $30 billion in aid, remittances and trade from the Arab countries is much larger than the oil import bill of the Third World countries. However, these compensatory payments should not be allowed to lessen the resolve of developing countries to curtail their consumption of increasingly scarce energy resources. They must be encouraged to plan ahead for the more efficient utilization of oil and for harnessing alternative energy supplies.

The Oil-Producing Countries[21]

Until the 1950s the international oil industry was dominated by the international oil companies. As Owais R. Succari described it:

> The history of the 'Majors' resembles greatly ancient Greek mythology, in which the destiny of the whole world was left in the hands of a few gods. These gods warred against each other and occasionally married one another in giant ceremonies. They sometimes demonstrated their authority and anger by punishing the people on earth who were supposed to obey unquestioningly the will of the gods. Taking the arbitrary despotism of these 'mighty gods' for granted, people didn't question the validity of their existence for a long time.[22]

However, the divine rule of the Majors has finally come to a halt. The international system has undergone great changes since the end of the Second World War. One of the most dramatic elements has been the renaissance of the Third World states. Even before the war, a new factor in world oil was making itself felt — the promising oil resources of the Middle East — and, as we saw earlier, special attention was paid to exploring that important area. In 1945 a special technical mission authorized by the US Secretary of the Interior to study the oil prospects of the Middle East declared that:

The center of gravity of world oil production is shifting from the Gulf-Caribbean area to the Middle East — to the Persian [Arabian] Gulf area — and is likely to continue to shift until it is firmly established in that area.[23]

As they came to realize the increased importance of the Middle East as the major oil-producing area and their capacity to influence the operating position of the long-established oil companies, the oil-producing countries of the Middle East began to challenge the injustices done them by existing oil concession agreements:

A feeling of discontent, generated by these regime's concessions bringing about the conviction that the status and obligations of foreign operators had to be subjected to a new order, rose high in many producing countries.[24]

After all, the history of oil is a history of imperialism. During the twentieth century, the fate of the oil-producing countries has been almost totally shaped by the forces of Western imperialism, particularly the international oil companies and their home governments. The large international companies have generally established themselves in the producing countries by highly questionable means. As described by Sheikh Mana Saeed al-Otaiba, the Oil Minister of the United Arab Emirates:

The large oil concessions, particularly in the Middle East, were granted under political and social conditions that were unfavourable for the producing countries, since most of them at the time did not exercise full sovereign rights over their territories or over their natural resources; moreover, the nationals of such countries lacked the knowledge and experience necessary for the proper evaluation of the real value of the oil rights granted.

Iraq, for instance, was threatened with the dismemberment of the Mosul Province from its territory unless the government agreed to grant a concession to the then Turkish Petroleum Company, later renamed Iraq Petroleum Company (IPC).[25]

It is no wonder, then, that in the Middle East countries the term 'concession' acquired an unfavorable connotation, 'so much so that at times its use has been condemned as derogatory to national honor'. Nevertheless, the development of the modern oil industry in the area began with this concept. Although these concession agreements differed from country to country, a number of characteristics were common to practically all concessions granted by the producing countries. The common elements can be summarized as follows:

(1) The area of concession was generally very large. In some cases it embraced the whole country (as in the case of Kuwait, Bahrain and Qatar); in others it covered a major region which, either because of its geological formation or because of its political character, constituted a well-defined territorial entity (as in the case of the Iranian concession, which excluded five northern provinces then under the Russian sphere of influence).

(2) The privileges granted to companies included, as a rule, the exclusive rights to explore, prospect, extract, refine and export crude oil and related materials (such as natural gas) within the area of the concession.

(3) The duration of the concession was specified, usually from 60 to 75 years.

(4) Companies were generally required to supply the oil requirements of the host governments and oil products for local consumption. This clause was usually accompanied by a proviso that quantities thus supplied should not be subject to royalty calculation and should be calculated at prices below those prevailing in world markets.

(5) Installation rights and the right of eminent domain were granted. Within these limitations, companies had the right to establish their own systems of transportation and communication for the efficient conduct of their operations. Among these facilities, radio, telegraph and telephone, as well as railroads, vessels and airplanes, were usually specified in the concession agreements.

(6) Certain extra-territorial rights, such as freedom from all direct and indirect taxation and freedom from government controls over the conditions of production and marketing, were granted.

(7) Companies were required to present to the host government an annual report of their operations, including data on the discovery of new oil deposits and geological plans and records. Such information was to be treated as confidential by the host country.

(8) The concession agreements made no mention of the surrender by the oil companies of unexploited areas after a certain period of time. Thus the companies were able to retain all the areas covered by the original concession agreement, although no actual operations were undertaken. On the other hand, the companies could refuse to allow

other companies or the host government to exploit these areas.

A United Nations report commented on the agreements made by the oil companies in the Middle East as follows:

The terms of their concessions give the foreign companies a freedom of action which substantially insulates them from the economy of Middle Eastern countries. Output is determined by considerations of world rather than local conditions. Moreover, it is the company which provides and owns the means of transport, whether pipelines or tankers, to carry Middle Eastern oil to its markets. The foreign exchange derived from sales of oil accrues to the petroleum companies and, in large measure, is retained by them. Hence, the impact of oil operations in Middle Eastern producing countries is mainly indirect and the benefits derived by them are limited.[26]

Commenting on the injustices of the oil concessions, George W. Stocking, the well-known oil economist, said: 'Never in modern times have governments granted so much to so few for so long.'[27]

Thus it should come as no surprise that, after attaining political independence, the oil-producing countries began to ask for major revisions of the terms of the old concessions. Between the early 1940s and the formation of OPEC, the host governments took actions ranging from direct negotiations between the governments and the oil companies for revision of concession agreements, to nationalization. However, the major breakthrough came only in 1960 with the creation of OPEC.

Notes

1. Peter R. Odell, *Oil and World Power* (Penguin Books, New York, 1979), pp. 9-14.

2. The seven 'International Majors' were until very recently responsible for something like 80 percent of oil production outside the communist countries and North America. However, since 1974 most of the oil-exporting producers have nationalized these companies' producing facilities. The companies, however, still retain control over the marketing of the oil produced in these countries. See ibid., p. 11.

3. The Standard Oil Trust was established by the Rockefellers in 1882 but was dissolved under anti-trust legislation in 1911. Since then its largest component, Standard Oil of New Jersey, has operated under a number of names. It was commonly known as Esso, but because of anti-trust limitations, it was also marketed in parts of the US as Humble Oil. In the early 1970s it tried to assume

one unified brand name and fixed on the word Enco, until it discovered that this was the Japanese word for stalled car. Finally it came up with Exxon. See Fred Halliday, *Arabia Without Sultans* (Penguin Books, Harmondsworth, 1974), pp. 395–6.

4. BP was originally known as the Anglo-Persian Oil Company (APOC) until 1933, when the country's name was changed. The company then became the Anglo-Iranian Oil Company (AIOC) and retained this designation until 1951, when Mossadeq nationalized it.

5. For a comprehensive account of these companies' history and activities, see Anthony Sampson, *The Seven Sisters* (Viking Press, New York, 1975) and Edith T. Penrose, *The Large International Firm in Developing Countries* (M.I.T. Press, Cambridge, Mass., 1968).

6. Enrico Mattei, *Problems of Energy and Hydrocarbons* (ENI, Rome, 1961), pp. 13ff. Quoted in Henry Madelin, *Oil and Politics* (Saxon House, D.C. Heath, London, 1974), p. 3.

7. Harvey O'Connor, *Empire of Oil* (Monthly Review Press, New York, 1962), p. 5.

8. Michael Tanzer, *The Energy Crisis: World Struggle for Power and Wealth* (Monthly Review Press, New York, 1974), p. 31.

9. Ahmed Zaki Yamani, 'Oil: Towards a new Producer – Consumer Relationship', *The World Today,* vol. 30 (November 1971), p. 480.

10. This approach was adopted by Michael Tanzer, who believed that putting oil-importing countries in one category ignored the essence of international oil relationships. The writer adopts this approach for a purely economic reason, but does not accept the tendency to categorize some of the Third World countries in a political grouping such as the 'Fourth World'. In the writer's opinion, such a trend seems to miss the essence of Third World politics. It is a concept which seeks to divide the developing countries between 'oil-haves' and 'oil-have-nots', a concept which has a far-reaching political objective: to divide the Third World countries' solidarity.

11. O'Connor, *Empire of Oil*, p. 10.

12. Quoted in L. Vernon Gibbs, *Oil and Peace* (Paxer, Stone & Baird, Los Angeles, Calif., 1929), pp. 33–4.

13. The Royal Dutch Company was organized in the Netherlands in 1890 for the purpose of developing a petroleum industry in the Dutch East Indies. In 1907 it affiliated with Shell Transport and Trading Company (Ltd.), which originally had carried mother-of-pearl shells in the Far East but had since become an important oil trader. The joint enterprise which evolved was called Royal Dutch–Shell, of which 60 percent was held by Dutch interests and 40 percent by British interests.

14. *Correspondence between His Majesty's Government and the United States Ambassador Respecting Economic Rights in Mandated Territories* (Cmd. 1226, 1921), p. 4.

15. 'United States Foreign Petroleum Policy', *State Department Paper*, 10 Feb. 1944, subcommittee for *Investigation of the National Defense Program.* Additional report of the Special Committee Investigating the National Defense Program, US, Congress, Senate, 78th Cong., 2nd sess. (Washington, 1944), Report No. 10, Part 15, *Report of Subcommittee Concerning Investigations Overseas*, p. 576.

16. According to this agreement the US government obtained for several American companies a 23.75 percent interest in IPC. The American participants were Standard of New York, Standard of New Jersey, Pan American Petroleum and Transport (Standard), Atlantic Refining (Standard) and Gulf Refining.

17. Mosley, *op cit.,* p. 48.

18. Jack Anderson, *Washington Exposé* (Public Affairs Press, Washington, D.C.), p. 202.

19. Tanzer, *Energy Crisis*, pp. 104–5.

20. *The Times of India*, 20 Feb. 1974. Quoted in S. Manoharan, *The Oil Crisis: End of an Era* (S. Chaud & Co. Ltd., New Delhi, 1974), p. 120.

21. Emphasis here will be placed only on Arab countries.

22. Owais R. Succari, *International Petroleum Market: Policy Confrontations of the Common Market and the Arab Countries* (Université Catholique de Louvain, Louvain, 1968), p. 51.

23. US Congress, Senate Special Committee Investigating Petroleum Resources, Wartime Petroleum Policies Under the Petroleum Administration for War, 19th Cong., 1st sess., 28–30 Nov. 1945, pp. 6 and 10.

24. Fuad Rouhani, *A History of OPEC* (Praeger, New York, 1971), p. 45.

25. Mana Saeed al-Otaiba, *OPEC and the Petroleum Industry* (John Wiley & Sons, New York, 1975), p. 25.

26. United Nations, *Review of Economic Conditions in the Middle East*, 4/1910 (New York, 1951).

27. George W. Stocking, *Middle East Oil: A Study in Political and Economic Controversy* (Vanderbilt University Press, Nashville, Tenn., 1970), p. 130.

3 OPEC: OIL CONSCIOUSNESS IN THE MIDDLE EAST

The creation of OPEC represented a major turning-point in the relationship between the producing countries on the one hand, and the oil companies and consuming countries on the other. It was the first attempt by the governments of oil-producing countries to organize themselves collectively.

Attempts by the oil-producing countries at co-ordinating their pricing and output policies go back to the mid-1940s, when the Arab League was formed. The League urged its members to co-ordinate their oil policies *vis-à-vis* foreign oil concessionaires, but the League's effectiveness was limited until the Arabs finally recognized that any sound oil policy must also take into consideration the non-Arab oil-producing countries. At the time, Venezuela was calling for a collective approach to confront the major oil companies because of their absolute domination over the international oil industry. In 1947 diplomatic contact was established between two of the major oil-producing countries, Venezuela and Iran, to achieve that goal. Later on, in September 1949, a special Venezuelan delegation was sent to the Middle East to promote co-operation among the oil-producers. This was followed in 1953 by the signing of the Saudi–Iraqi agreement signalling an important development within the Arab world. The two governments called for the exchange of oil information and the holding of periodic consultations on mutual policies to be agreed between the two countries. This was the first OPEC-type grouping between oil-producing countries to provide actual co-operation at governmental level in the oilfields. However, concrete steps to broaden the agreement towards the establishment of an association of oil producers did not materialize until the late 1950s.

It is now generally recognized that the successive, unilateral cuts in oil prices by the major oil companies in 1959 and 1960 were directly responsible for the creation of OPEC. The initial price reduction in 1959 provided the spur to unity. During April of that year, the Economic Council of the Arab League sponsored the First Arab Petroleum Congress in Cairo. Iran and Venezuela were invited to attend as observers. The Congress recommended that there should be no reduction in the posted price by the companies without prior consultation with the oil-producing governments. Arabs and non-Arabs exchanged views concerning the creation of an organization of oil-exporting countries. Moreover, private talks were held between Abdullah Tariki, Saudi Oil Minister and Pérez Alfonso, his Venezuelan counterpart, which led to a secret gentlemen's agreement; this, according to Alfonso, 'constituted the first seed of the creation of OPEC'.[1]

Despite the strong position taken by the oil-producing countries and their warning to the companies against a unilateral price reduction, the companies went ahead and, for the second time, cut oil prices in August 1960. This action reduced Middle East posted prices to a level below that of 1953. The two price cuts of 1959 and 1960, of approximately 27 cents a barrel, resulted in a drastic loss of revenue to the major Middle Eastern oil-producing countries, amounting to an estimated loss of $4 billion to the decade 1960–70.[2]

The five countries responsible for 80 percent of the world's exports of oil – Saudi Arabia, Iran, Iraq, Kuwait and Venezuela – met in Baghdad from 10 to 14 August 1960, and on 15 September announced the establishment of the Organization of Petroleum Exporting Countries (OPEC). The principles governing the organization are clearly stated in paragraph (4) of Resolution 1.2, adopted by the founding members at their conference in Baghdad, and stipulate 'the resolution of petroleum policies for the Member Countries and the determination of the best means for safeguarding the interests of Member Countries individually and collectively'. The resolutions also make clear that the oil producers' chief opponents are the oil companies, and go on to declare:

That members can no longer remain indifferent to the attitude heretofore adopted by the oil companies in effecting price modifications; that members shall demand that oil companies maintain their prices steady and free from all unnecessary fluctuations;
that members shall endeavor, by all means available to them, to restore present prices to the levels prevailing before the reduction;
that they shall ensure that if any new circumstances arise that in the estimation of the oil companies necessitate price modifications, the said companies shall enter into consultation with the member or members affected in order fully to explain the circumstances . . .

As defined by Fuad Rouhani, the first Secretary-General of OPEC, the aims of the organization are as follows:

(1) Endeavoring by all possible means to restore the price of crude oil to the level existing before the reductions.

(2) Ensuring that the oil companies maintain their prices at a stable level, avoiding any unnecessary fluctuations.

(3) Demanding that the companies should consult the countries concerned each time they consider that the position of the market calls for a modification of prices; in such an event, they would have to furnish a full explanation to justify their point of view.

(4) Elaborating a formula to ensure the stability of prices, and for this purpose resorting, if necessary, to a regime of control of production.

(5) Conducting the Organization's studies and activities in conformity with the following objectives: guaranteeing a stable revenue to the producing countries, an effective, economic and regular supply of petroleum to consumers, and an equitable remuneration to investors.

(6) Organizing a system of regular consultation among the members with a view to co-ordinating and unifying their petroleum policies, and determining the best means of safeguarding their common interests, both individually and collectively.[3]

Thus OPEC was established as a 'defensive' body to form a common front *vis-à-vis* major expatriate oil firms and major oil-importing countries. OPEC's aims were limited to the co-ordination of members' export policies in one commodity, oil. The Arab members of OPEC did not broaden their concerns to include regional integration of members' economies and societies until they established OAPEC in 1968.

The Organizational Structure of OPEC

In January 1961 agreement was reached on the basic organizational structure of OPEC, and was embodied in the Caracas Declaration. Three main bodies were set up: the Conference, the Board of Governors and the Secretariat (see Figure 3.1).

The Conference

The Conference is regarded as the supreme authority of the organization. It is comprised of one delegate from each member country and meets at least twice a year. The Conference formulates general policies and identifies how they will be implemented; decides on applications

Figure 3.1: OPEC's Organizational Structure

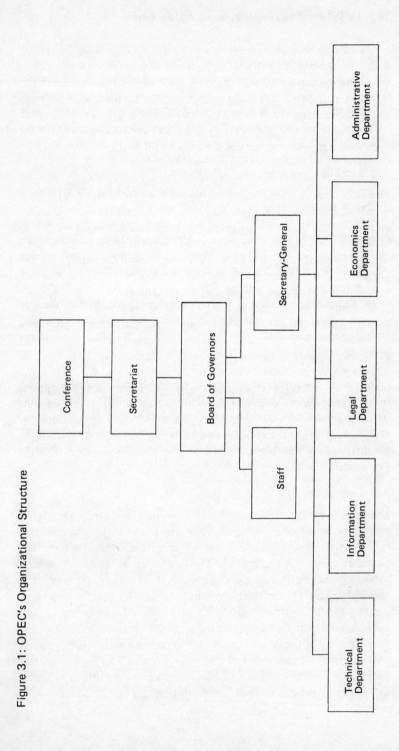

for membership; confirms the appointment of Governors; directs the Board of Governors to submit reports and recommendations and considers them when submitted; approves the budget for each year at the recommendation of the Board, and approves the liquidation of each year's accounts on the basis of auditors' reports; appoints the Chairman of the Board, the Secretary-General, his Deputy, and the Auditor for each year; and approves any amendments to the Statutes.

The Board of Governors

The Board of Governors includes one representative of each member country, nominated for two years by their respective governments and confirmed by the Conference. The Board holds at least two meetings each year, and may hold an extraordinary meeting at the request of the Chairman, of the Secretary-General or of two-thirds of the Governors. The Chairman of the Board is designated for a one-year term by the Conference from among the Governors.

The functions of the Board are as follows: to control the management of the organization's affairs and the execution of the Conference's decisions; to deliberate and decide on reports submitted by the Secretary-General; to submit reports and recommendations to the Conference on any subject of interest to the organization; to draw up the budget for each calendar year and nominate an Auditor for approval by the Conference; to consider the Auditor's Report for each year prior to its submission to the Conference; to approve the appointment of Chiefs of Departments designated by the Secretary-General upon nominations by member countries; and to prepare the agenda of the Conference.

The Secretariat

The Secretariat is permanently based in Vienna, the seat of the organization. It is entrusted with carrying out the executive functions of OPEC under the general direction of the Board of Governors. The Secretary-General is the legally authorized representative of the organization, appointed by the Conference for three years, and must be a national of one of the member countries. Until 1965 the Secretary-General was concurrently Chairman of the Board of Governors. This dual capacity was abolished by the Eighth Conference; from May 1965 the Chairman of the Board has been nominated separately each year.

The functions assigned to the Secretary-General are: to organize and administer the work of the organization; to ensure the proper functioning of the Secretariat's Departments; to prepare, for the Board, reports on studies requested by the Conference and on the implementation of decisions of the Conference; and in general to carry out the instructions of the Conference. These functions are carried out by the office of the Secretary-General and five Departments: Technical, Administration, Legal, Economics and Information. The Secretary-General may, when the need arises, call for legal and other professional advice from outside the Secretariat, and may set up *ad hoc* working parties.

The founding members in 1960 included: Iran, Iraq, Kuwait, Saudi Arabia and Venezuela. Qatar joined the organization in 1961, followed by Indonesia and Libya in 1962, Abu Dhabi in 1963, Algeria in 1969, Nigeria in 1971 and Ecuador in 1973. Gabon joined as an associate member in 1974; and in that year Abu Dhabi was replaced by the United Arab Emirates.

In 1980 OPEC controlled over 67 percent of the world's proven oil reserves and 32.8 percent of the world's gas reserves. In the same year, OPEC produced 45 percent of world oil output, and accounted for over 85 percent of international trade in oil. These percentages confirm the central role of OPEC in the world energy picture.

Notes

1. Pérez Alfonso, 'The Organization of Petroleum Exporting Countries', *Monthly Bulletin*, no. 2 (1966).

2. Mana Saeed al-Otaiba, *OPEC and the Petroleum Industry* (John Wiley & Sons, New York, 1975), p. 53.

3. Fuad Rouhani, *A History of OPEC* (Praeger, New York, 1971), p. 79.

4 OPEC'S ACHIEVEMENTS

OPEC's achievements over its 20 years of existence may be evaluated in terms of the net benefits member governments have received, whether economic or non-economic. However, to judge its success or failure, one has to understand the main purposes which OPEC proclaimed for itself. OPEC concentrated on establishing a uniform policy, 'ironing out' the differences that existed between the oil-producing countries and the oil companies with respect to prices, royalties and production. During the first ten years of its existence, OPEC did not reach these goals, but since 1970 its has been increasingly effective.

OPEC and the Price of Oil

The low price of oil can be regarded as the primary cause of the creation of OPEC. As J.E. Hartshorn observed:

> The Organization of Petroleum Exporting Countries did not come out the blue . . . It came as a direct reaction . . . to one of the rather infrequent changes in price in this industry — a reduction, just about a month before, in the 'posted price' of crude oil in the Persian Gulf . . . The attitudes that were finally expressed in the formation of OPEC had begun to form many years before, too. But it took the price cuts of August, 1960 — which may turn out to have been the last that these companies were ever to make in these prices, in their then form — to crystallize them.
>   . . . Without the price cuts that Standard initiated . . . the Organization of Petroleum Exporting Countries [OPEC] would almost certainly not have come into being — at the time.[1]

The first declaration by OPEC emphasized the question of prices. Thus the first task which OPEC set for itself was to stabilize oil prices and keep them 'steady and free from all unnecessary fluctuations'. Accordingly, the first paragraph of the first resolution asserted that 'members can no longer remain indifferent to the attitude theretofore adopted by the oil companies in effecting price modifications'. OPEC's initial declaration on price restoration was not followed up with a

38

concrete policy to achieve that objective until two years later at the Fourth OPEC Conference. However, OPEC's collective action during that time was not yet developed and the members achieved very little individually.[2] Nevertheless, the organization succeeded, to some extent, in preventing the oil companies from imposing further reductions in oil prices. In the words of Joe Stork:

> Even though OPEC failed in its endeavor to restore posted prices, and in effect helped to diffuse the popular political momentum that had led to its creation, the organization, merely by its continued existence and potential power, did succeed in preventing any further cuts in the face of the continually declining oil prices on the European market.[3]

The OPEC members finally realized that they had insufficient power to force oil prices up, and from 1962 on, they sought to obtain by means of increasing royalties what they had been unable to obtain by demanding price increases.

OPEC and Oil Royalties

Under the conditions of the earlier concession agreements, royalties were the producing countries' major source of income from oil production. The concept of royalties developed with the evolution of the oil concession. In some of these agreements, royalties were defined as a fixed sum per ton of production, sometimes as a percentage of the concessionaire's profit (for example, 16 percent under the terms of the old D'Arcy concession in Iran).

After failing to press successfully for any change in oil prices, OPEC launched phase two of the campaign to secure better arrangements for the producing countries. First, OPEC eliminated various allowances and discounts by which the large oil companies secured greater profits; and, secondly, they improved the 'take' of the oil-producing governments from royalties.

At the time OPEC was created, royalty payments to host governments were in the form of royalty per barrel, usually 12.5 percent of the posted price. The only OPEC member country that benefited from a rate higher than 12.5 percent was Venezuela.[4] In addition, the companies treated royalty payments as a necessary production cost in calculating profits. When the companies agreed on the 50/50 profit-sharing formula in the early 1950s they calculated the royalties as part of the government's 50 percent, thereby making the government's actual share much less.[5] The royalty, in effect, became a payment to the companies rather than the countries. In the words of Fuad Rouhani, OPEC's chief negotiator with the oil companies:

> Either the . . . companies are paying income tax at the full rate prescribed by law, but no royalties, or they are effectively paying royalty but their income tax payment amounted to about 41% of income, not 50%.[6]

Against the foregoing background, Resolution 33, adopted at OPEC's Fourth Conference in Geneva in 1962, demanded:

> That each member country affected should approach the company or companies concerned with a view to working out a formula whereunder royalty payments shall be fixed at a uniform rate that members consider equitable, and shall not be treated as a credit against income tax liability.

However, approaching the companies to discuss the issue of royalties was not an easy task. The negotiations on this issue, which lasted from 1962 to 1965, have been characterized by OPEC as the 'longest, toughest and most revealing in the history of the international oil industry'. Throughout the negotiations, the oil companies persistently rejected the principle of collective bargaining, which the OPEC countries adopted to deal with the companies. Although two or more of the companies were joint owners of concessions in each of the Middle Eastern exporting countries, and were accustomed to using their collective power in negotiating with individual countries, they were reluctant to recognize and accord a similar joint interest to the countries with which they dealt. *The Times* referred to this action of the oil companies in an editorial:

> This [opposition] could be represented as yet another instance of the [oil companies'] attempts to slow down the whole process of negotiation and to behave in an out-of-date imperialistic way. To some extent the industry would have only itself to blame if this happened. Some of its more prominent members have not hesitated to treat OPEC as an inevitably hostile body — if indeed, they are prepared to recognize its existence at all. Yet the Governments have

a perfect right to set up a body to look after their common interests. Moreover, these interests are to a large extent shared with the companies. All make energy out of oil. If producing governments sometimes forget this, the industry should show its great wisdom by being less haughty about OPEC.[7]

Eventually, the companies reluctantly agreed to bargain with OPEC. The protracted negotiations ended in the companies' agreeing to improve the royalties of the oil-producing countries; in other words, royalties were to be deducted before profits were calculated and divided, thus increasing the actual government 'take' per barrel of crude. However, the royalty settlement permitted the companies to discount the posted prices by 8.5 percent in 1964, 7.5 percent in 1965, and 6.4 percent in 1966, in calculating their income tax obligations.[8] The settlement provided, moreover, for consultation in 1966 between the governments and the companies on possible future reductions in the discount rate. In April 1966, at OPEC's Eleventh Conference in Vienna, the member states adopted Resolution X171, advising member countries to 'take steps towards the complete elimination of the discount allowance granted to the oil companies'.

In accordance with this recommendation, negotiations soon reopened between the producing countries and the companies. However, as usual, negotiations proved difficult, and there was no solution yet in sight when the Arab-Israeli war broke out in 1967. Following the outbreak of war, the Arab countries decided to boycott oil shipments to certain countries; the boycott, together with the closure of the Suez Canal, caused crude oil prices to recover. The interruption of oil supplies and the firming of the oil market improved OPEC's bargaining position. At the OPEC Conference in Rome in September 1967, the members decided that the five countries most concerned with expanding royalties (Saudi Arabia, Iran, Kuwait, Libya and Qatar) should meet for consultation on the issue in early October. On 9 January 1968, after a two-day conference in Beirut, the five OPEC members announced that they had accepted an offer submitted by the companies on 6 January to Iran and Saudi Arabia, binding all companies operating in OPEC member countries. Under the agreement reached, the discounts were to be phased out over a four-year period, declining from 5.5. percent in 1968, to 4.5 percent in 1969, 3.5 percent in 1970, 2 percent in 1971, and ceasing entirely in 1972. The stage was set for the October oil 'revolution' of 1973.

OPEC and Control of Production

Since 1960 OPEC has examined the possibility of developing a system of overall production control. The protagonists of this idea in the Arab bloc was Sheikh Abdullah Tariki, then Oil Minister in Saudi Arabia. Tariki discussed this idea with Pérez Alfonso, the then Venezuelan Minister of Mines and Hydrocarbons, during the First Arab Petroleum Congress in April 1959. A fresh exchange of views on this subject took place between the two ministers when they met at the Fifth World Petroleum Congress in June 1959 in New York. Although they realized the difficulties inherent in such a scheme, both men were persistent in their belief through co-operation, the oil-producing countries would work out a formula to achieve their common objectives.

At the Second Arab Petroleum Conference held in Caracas on 16 January 1961, OPEC acknowledged that a 'just-pricing formula' supported by international pro-rating would require detailed study. An American company, Arthur D. Little, Inc., was hired to make a comprehensive analysis of profits, prices and output in the international oil industry and to put forward recommendations for increasing member countries' control over production. Although the recommendations made by Arthur D. Little were never made public, it is widely known that the study, for obvious reasons, was not very encouraging to OPEC members. However, the oil-producing countries did not abandon their plans.

Even before OPEC was founded, national companies were set up in many of these countries to participate in and monitor oil production. Some of these companies include: the National Iranian Oil Company (NIOC) set up in Iran, 1954; the Kuwait National Petroleum Company (KNPC) in Kuwait, 1960; Petromin in Saudi Arabia, 1962; Sonatrach in Algeria, 1963; the Iraq National Oil Company (INOC) in Iraq, 1965; and the Libyan National Oil Company (LINOCO) in Libya, 1969. Their explicit aim was to take over a larger share of oil-generated rents. In many cases, after 1960, these national companies took advantage of the Newcomers' desire to take up concessions, to sign contracts with them that gave the national companies some control over oil operations in their countries.

OPEC's Success in the 1960s: an Assessment

Until 1970 the results obtained by OPEC were limited in scope. In general, it may be said that OPEC's greatest success in the 1960s came in areas where the large companies had traditionally been most willing to make concessions to the individual governments; conversely, where their demands most threatened the vital interests of the companies, OPEC had virtually no success. However, despite its limited impact, OPEC had been useful to its members; the oil-producing countries would have been worse off without it. Moreover, OPEC demonstrated that concerted action by Middle Eastern governments was feasible; in the words of Don Peretz: '[The] continuation and flourishing of OPEC for over a decade has undermined one of the prevailing stereotypes about Middle Easterners: that they can never accomplish anything together.'[9]

OPEC's basic inability to obtain fundamental changes in the position of the oil-exporting governments in the 1960s, and its failure to achieve its principal objectives *vis-à-vis* the major oil companies, can be attributed to several factors.

First, the major oil companies did not readily concede any point where their vital interests, particularly control over prices or production, were at stake. In effect, they used all possible means, especially legal procedures, to retain control over production.

Second, attempts by the producing nations to gain control over their national resources met with opposition from the principal consuming countries. Any attempt by OPEC members to control oil production was considered illegal and a direct interference with the free play of supply and demand.[10]

Third, countries just beginning to export crude, like Libya, Algeria, Nigeria and Abu Dhabi, were not in a position to agree to restrict the growth of their production, for restrictions would have led to an immediate loss of major revenues. At the time, they considered it more advantageous to increase the level of their total revenues by selling ever-increasing quantities, even at a relatively low price, rather than to restrict their production.

Fourth, OPEC's great emphasis on negotiation and cultivating favorable world opinion was interpreted by both companies and consumers as a sign of weakness. Negotiations were allowed to drag on, in a way which left the major companies with the initiative and the producing countries with the option only of accepting or rejecting company offers:

In appealing to world opinion, OPEC runs essentially the same risk as it does in appealing to a moderate body of opinion in the Arab world itself. That is to say, it inevitably strikes a note of patient exhortation, one might almost say entreaty, which tends to jar on those more nationalist in outlook, who believe that the Arabs and all in their situation should adopt forceful methods as a matter of sovereign right.[11]

Abdullah Tariki, a co-founder of OPEC and Saudi Arabia's Director-General of Petroleum and Mineral Resources in 1960, sharply criticized the organization's achievement in 1965 on the same grounds. Member countries should not, he said, follow the 'soft' path of negotiation and compromise; rather, they should assume the prerogatives of sovereignty by legislating simultaneously if positive results were to be obtained and expectations to be fully realized. In an impassioned speech at the Fifth Arab Petroleum Congress he argued:

> The OPEC delegates had better fasten their seat belts; for the fact is that OPEC deserves little or no credit . . . In the [1964] negotiations with OPEC member governments, the companies agreed to the principle of royalty expensing but only on the condition that they were given an 8.5% discount off posted prices. What a humiliation for the companies to dictate to a group of sovereign governments in this way – particularly considering that OPEC's main aim was the restoration of posted prices to the pre-August level![12]

OPEC had no real negotiating strategy beyond repeated declarations of its 'sense of responsibility and good faith'; it stopped short of 'being a menace and a threat to the security of the international oil industry'.

OPEC's weakness seemed to reflect basic divisions among OPEC member countries. These differences prevented them, in the early 1960s, from creating a united front against the oil companies. Force and unity would prove to be the only language the companies would understand. As Michael Tanzer puts it: 'The companies [and their home governments alike] would only yield where there was a real danger of being forced into a worse situation if they did not.'[13]

Nevertheless, after more than a century of subservience to the joint oil companies and their Western home governments, rising oil consciousness in the Middle East and in the other Third World oil-producing areas had begun to undermine the foundations of century-long domination. The post-war period, particularly the 1960s, severely limited the freedom of the international oil companies to make decisions solely in the light of their own corporate interests and to follow up opportunities for expanding their activities in refining and marketing oil. In the exporting countries the awakening was inescapable, but it came slowly because the Majors possessed powerful means of delaying it.[14] The producing countries moved ahead less rapidly than they desired and it was not until the 1970s that a major shift of power was achieved:

> To conclude, it could be said that, during this period, the balance of power scarcely altered. In 1970 they [the oil companies] still controlled about 80 percent of world crude exports and 90 percent of Middle East production. For these companies the Middle East was not only a fabulous source of wealth, but also a means of adjusting supply to demand, since it was sufficient to open or close the taps to increase or reduce production depending on requirements . . . Until 1970 consumers had been supplied at relatively secure prices with few problems . . . The Suez crisis was a grave warning . . . The Six-Day war and the closure of the Suez Canal in 1967 had very little effect on supplies to Western Europe. [However] this regularity of supply and the stability of prices were to be suddenly threatened by the [October] crisis.[15]

The Triumph of OPEC

After 1970, within the space of a few short years, OPEC was able to topple the old oil regime whereby the oil companies dictated production rates, prices and revenues to each oil-producing country. The new order that OPEC created restored control to the legitimate owners of the oil reserves and empowered them to make independent decisions within the framework of burgeoning national sovereignty, co-operative action within OPEC and an emerging world role.

OPEC countries finally emerged as the sole determinants of their oil policies, but not without a major effort, as Chapter 11 reveals. This rise to power brought in its wake enormous dividends. The price of oil quadrupled between 1972 and 1974, and doubled again between 1979 and 1980 (see Figure 4.1). Oil revenues increased substantially, as is shown by Table 4.1, which presents oil revenues for each member of OPEC and total group revenues from 1961 to 1980. Between the discovery of oil in the developing countries and 1973, almost one third of their total proven oil reserves were used up with little benefit to these countries. The turning-point in 1973 has allowed these countries the

Figure 4.1: Oil Prices, 1971–1981 (US dollars per barrel)

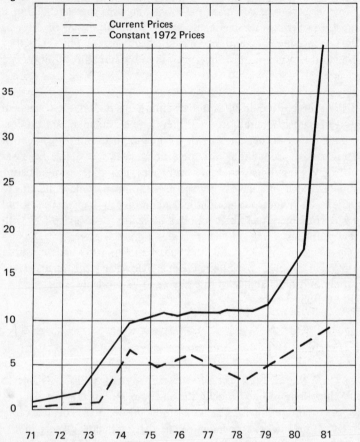

opportunity to use oil as an instrument of economic and social development. All OPEC members face the same challenge: to stretch the exploitation period of their oil reserves long enough to enable them to establish a productive economic system capable of replacing oil as a source of income and export proceeds, in order to sustain comparable standards of living to those they are now at long last beginning to enjoy.

As long as oil is the only engine for growth, and it is likely to remain so within the foreseeable future, OPEC members will have to co-operate and harmonize their policies to sustain their collective benefits. Together they have a chance to succeed in meeting their common objectives. Divided they will fail and suffer the consequences.

Table 4.1: OPEC Member Countries' Oil Revenues, 1961–1980 (millions of U.S. dollars)

	1961	1962	1963	1964	1965	1966	1967	1968	1969	1970
Algeria	28.0	45.0	52.0	60.0	76.0	128.0	199.1	251.8	266.7	271.9
Ecuador	.4	.4	.4	.6	.8	.8	2.5	3.5	4.0	4.3
Gabon	—	—	—	—	—	—	—	—	—	—
Indonesia	96.2*	96.3*	102.0*	97.0*	97.9*	97.4*	109.0*	133.1	201.8	254.3
Islamic Rep. of Iran	291.2	342.3	380.0	482.2	514.1	608.2	751.6	853.4	922.8	1109.3
Iraq	265.4	266.3	308.0	353.1	367.9	394.2	364.4	487.9	479.0	521.0
Kuwait	461.4	480.3	521.4	566.4	598.2	637.8	714.7	701.8	759.1	820.7
S.P. Libyan A.J.	3.0	39.8	107.8	210.6	351.1	522.8	625.0	1001.8	1175.2	1351.3
Nigeria	18.7	23.8	16.5	20.3	36.3	57.5	59.6	45.7	89.5	247.0
Qatar	54.3	56.6	60.5	63.8	69.2	90.7	105.4	111.8	117.6	122.4
Saudi Arabia	377.6	409.7	455.2	523.2	664.1	789.9	903.6	926.4	949.2	1214.0
United Arab Emirates	—	2.0	6.4	12.3	33.3	99.7	109.6	153.4	190.8	233.0
Venezuela	842.7	962.8	994.4	1080.3	1097.1	1074.8	1213.3	1223.0	1228.6	1376.8
Total OPEC	2439.4	2725.4	3004.6	3469.8	3906.0	4501.8	5157.8	5903.6	6384.3	7526.0

	1971	1972	1973	1974	1975	1976	1977	1978	1979	1980
Algeria	321.0	613.3	987.7	3299.2	3261.8	3699.0	4253.7	4589.1	7513.0R	10787.0
Ecuador	—	29.7	128.8	413.9	292.8	532.7	499.3	500.0	800.0	1200.0
Gabon	9.4	18.1	29.4	172.7	483.5*	800.0	600.0	600.0	900.0	1600.0
Indonesia	336.2	505.8	687.7	1364.3	3233.1	4465.7	4692.3	5200.0	8100.0R	10500.0
Islamic Rep. of Iran	1851.1	2396.0	4599.2	17821.8	18433.2	22042.9	21210.2	19300.0	20500.0	11600.0
Iraq	840.0	575.0	1843.0	5700.0	7500.0	8500.0	9631.0	10200.0	21291.0R	25981.0
Kuwait	954.3	1403.7	1734.7	6542.6	6393.1	6869.5	7515.7	7699.5	16863.0R	18016.0
S.P. Libyan A.J.	1674.2	1562.6	2222.9	5999.0	5101.0	7500.0	8850.0	8400.0	15223.0R	22527.0
Nigeria	846.8	1117.2	2048.0	6654.2	7422.1	7715.0	9600.0	7900.0	15900.0	20000.0
Qatar	199.6	255.2	463.1	1849.0	1684.9	2091.9	1994.0	2200.0	3642.0R	5377.0
Saudi Arabia	1884.9	2744.6	4340.0	22573.5	25675.8	30754.9	36538.4R	32233.8	57522.0	102212.0
United Arab Emirates	431.0	551.0	900.0	5536.0	6000.0	7000.0	9030.0	8200.0	12862.0R	19344.0
Venezuela	1674.7	1901.6	3028.7	9270.6	6968.0	7712.8R	8106.2R	7319.0R	11647.1R	14881.1
Total OPEC	11023.2	13673.8	22813.2	87196.8	92449.3	107884.8	122520.8	114341.4	192763.1	264025.1

* Estimated
R Revised.
Source: OPEC. Annual Statistical Bulletin 1980, p. XLIX.

Notes

1. J.E. Hartshorn, *Politics and World Oil Economics: An Account of the International Oil Industry in its Political Environment*, 2nd edn (Praeger, New York, 1962), pp. 20 and 23.

2. The inability of OPEC to produce any change in oil prices in the early 1960s could be attributed to various factors: the entry of Newcomers, the competition of Soviet crude, the appearance of new producing countries, the relative decrease in demand for oil in Europe, the restrictions on imports of the United States and, finally, the declining trend in costs, which made it nearly impossible to enforce joint action with an eye towards increasing prices.

3. Joe Stork, *Middle East Oil and the Energy Crisis* (Monthly Review Press, New York, 1975), p. 95.

4. The applicable rate of Venezuela was 16 2/3 percent.

5. At a posted price for crude of $1.80 a barrel with a 12.5 percent royalty (the equivalent of $0.255 a barrel) and allowed production cost of $0.20 a barrel, the government 'take' share would be (50/50 profit) calculated as follows:
(a) Royalty is credited against income tax: the sum due for one barrel is $1.80 minus $0.20 = $1.60. Thus the government share is $0.80.
(b) Royalty is expensed like other costs of production: the sum due for one barrel is $1.80 minus $0.20 minus $0.225 (12.5 percent of $1.80) = $1.375; government share is $0.6875 (based on 50/50 profit-sharing). Total government revenue will be $0.6875 (income tax) plus $0.225 (royalty) = $0.9125.

6. Quoted in Muhamad A. Mughrabey, *Permanent Sovereignty Over Oil Resources: A Study of Middle East Oil Concessions and Legal Changes* (Middle East Research and Publishing Center, Beirut, 1968), pp. 141-2.

7. *The Times*, 3 Aug. 1963.

8. Fuad Rouhani, *A History of OPEC* (Praeger, New York, 1971), pp. 228-9.

9. Don Peretz, 'Energy: Israelis, Arabs and Iranians' in Joseph E. Szyliowicz and Bard E. O'Neill (eds.), *The Energy Crisis and US Foreign Policy* (Praeger, New York, 1975), p. 91.

10. Rouhani, *History of OPEC*, p. 213.

11. Michael Tanzer, *The Energy Crisis: World Struggle for Power and Wealth* (Monthly Review Press, New York, 1974), p. 69.

12. *Middle East Economic Survey (MEES)*, 26 March 1965, p. 25.

13. Tanzer, *Energy Crisis*, p. 68.

14. The United States, using the CIA, financed a coup in Iran against the elected government of the national premier Dr Mohammad Mossadeq in 1953, thus abolishing the nationalization act which had been approved by the Iranian parliament in 1951. Similarly, it is alleged that oil companies financed the military take-over of the democratic regime of Venezuela in 1948. Colonel Perez Jimenez acceded to the companies' 50/50 profit-sharing rule. See Jean-Marie Chevalier, *The New Oil Stakes* (Penguin Books, London, 1975), p. 156.

15. Ibid., pp. 32-3.

PART TWO
ARAB OIL AND THE PALESTINE CONFLICT

5 OIL AND THE PALESTINIAN QUESTION

The oil crisis[1] is rooted in the broader issue of the Arab-Israeli conflict. This conflict is in turn an outgrowth of the more fundamental Palestinian question. Any attempt to resolve the oil crisis in isolation from the Palestinian question is doomed to failure because these two issues are inseparable. Resolution of the oil crisis is possible only as a by-product of an end to the Palestinian problem.

The Palestinian question is the fundamental cause of oil being used as a political instrument. Since the politicization of oil was mainly directed towards the West, it is necessary to explain the Arabs' perception of the West's role in the creation of the Zionist state in their midst and the subsequent displacement of Palestinians and other Arabs from their homelands. The significance of a historical perspective is eloquently summarized by Mohamed Heikal, former Editor-in-Chief of the Egyptian daily *Al-Ahram*, in his book *The Road to Ramadan*:

> The [Ramadan] war emphasized once again the importance of the historical dimension . . . Israelis and Americans have always been at fault in approaching situations in a strictly pragmatic way. They have dealt only with what they could see, concentrating on the present to the almost total exclusion of the past. How often in talks with Rogers, Kissinger, Sisco and others has Egypt heard the Americans say in effect, 'We're not interested in raking over the past; let's look at the situation as it is today.' But today's situation is the creation of yesterday.[2]

The roots of the Palestinian question go back to 1897, when the Zionist movement took political shape during the First World Zionist Congress in Basle, Switzerland. It was then that political Zionism first entered the dictionary of Middle Eastern politics. Although we do not intend to present a detailed history of the Arab-Israeli conflict, it is necessary to discuss briefly the historical aspects of the conflict as a background to the current oil crisis.

According to Arab nationalist writers, the history of the Arab-Israeli conflict can be divided into four major stages: the Infiltration Stage from 1882 to 1917, the Penetration Stage from 1917 to 1948, the Invasion Stage from 1948 to 1967 and the Expansion Stage from

51

1967 to 1973.[3]

The earliest waves of Jewish immigration to Palestine started as early as 1882 as a consequence of anti-Semitism[4] in Russia and Eastern Europe. This initial immigration led to the creation of the organization called *Hoveve Zion* (Lovers of Zion) in Odessa and the establishment of the first Zionist colonies in Palestine. The beginnings of immigration of Jews from Eastern Europe coincided with the British occupation of Egypt in 1882, and Arab nationalists have identified Zionism ever since as a form of Western imperialism in the Arab world.

The Zionist movement, however, remained largely in the backround until the 1890s, when Theodor Herzl (1860-1904) became its leader. His book *Der Judenstaat* (The Jewish State), published in 1896, gave the Zionist movement a definite political turn, and resulted in the earliest organized claim of Jews to Palestine in 1897. In that year, at Herzl's initiative, the First World Zionist Congress assembled in Basle, attended by some 200 delegates. Herzl, chosen as the first leader of the World Zionist Organization (WZO), conducted an elaborate effort to claim Palestine from the Ottoman Sultan, who was then in control of most of the Arab world. However, the Sultan rejected Herzl's effort.

Zionism also encountered some opposition among Jews themselves. However, despite initial setbacks, Zionism as a political movement continued to advance successfully. The Jewish settlement in Palestine increased, and by the outbreak of the First World War, there were some 56,700 Jews in Palestine. Most of them had emigrated from Eastern Europe and Russia. By 1917, the total Jewish population represented less than eight percent of the total population of Palestine. The success of the Zionist movement in bringing Jews to Palestine would have been limited had the British mandate authorities been unco-operative, but the British authorities in Palestine accorded the Jews, particularly until 1939, substantial aid and encouragement. The United States, which was hesitant about aiding Zionists before the Second World War, became heavily involved in securing their grip over Palestine after the war as British support waned, and thus helped to bring about the creation of the State of Israel in 1948.

It took half a century of Zionist-Western co-operation finally to establish the State of Israel in the midst of the Arab world. As a Western creation, Israel was inescapably regarded as a form of Western imperialism, a vanguard state to secure Western interests in the Arab world, necessary to the modern industrial revolution; and Israel would be in a strategic position to ensure the continued flow of oil from the Middle East by suppressing Arab attempts at national control of oil.

Oil interests were predominant among the factors moving Western countries to vote for the creation of Israel in 1948. The right of the Palestinian people to their homeland was sacrificed to the West's oil needs. Nahum Goldmann, the President of the World Zionist Organization, in addressing the WZO on 31 May 1949, spoke of a conversation he had had with Ernest Bevin, when British Foreign Secretary. According to Goldmann, Bevin stated:

Palestine today is the center of world strategy and the tool by which to control world politics. However, issues of justice or honor are not what the world is concerned with: people and governments define their positions according to their real interests . . . When a statesman or a politician thinks of Zionism today, he never thinks in terms of pragmatic facts: He directs his attention to oil, the Middle East, Russia, and the United States.[5]

Today, the Arab-Israeli conflict is a reflection of the power struggle between imperialism and nationalism. Western strategy involves a search for regional power bases which can serve Western interests around the world. In this strategy, Israel serves as a 'watch-dog' for Western interests in the Middle East. The 1956, 1967, 1978 and 1982 conflicts are vivid illustrations of this design. However, as this study will demonstrate, Israel is actually a liability to Western interests in the Middle East. With every major conflict, Western support for Israel has resulted in a major decrease of Western influence in the area. Although the Americans played a stronger role in the Arab world after the 1973 war, American differences with the Arabs over Israel in fact have become sharper. It is likely therefore that the apparent strengthening of American influence in the Arab world in the 1970s is precarious and likely to be short-lived.

Notes

1. By oil crisis, we mean the political use of oil as an instrument to effect Arab objectives. We are referring here to the simple economics of imbalance between demand and supply.
2. Mohamed H. Heikal, *The Road to Ramadan* (Quadrangle Books, New York, 1975), pp. 260–1.
3. Ahmed S. al-Dajani, *Maza Ba'd Harb Ramadan?* (What After the Ramadan War?) (United Publishing Co., Beirut, 1974), p. 57.
4. The conventional term 'anti-Semitism' is not technically correct,

particularly when it is used in the context of the Arab–Israeli conflict, as the Arabs are themselves Semites.

5. Quoted in Hamed A. Rabie, *Al-Ta'awum al-Arabi wa al-Siyasah al-Batroliyyah* (Arab Co-operation and Oil Policy) (Cairo Modern Library, Cairo, 1971), p. 18.

6 THE WEST AND THE CREATION OF ISRAEL

To the Arabs, Zionism is one form of imperialism. They perceived it to be a threat to their very existence even when it was still but an ideology. They followed with intense concern its emergence as a political force, and the attempts by Zionists to establish relationships with world powers in the late nineteenth century.

The Arabs were aware of the early Zionists' efforts to gain a foothold in Ottoman-dominated Palestine. Although the Ottoman Sultan refused at the time to yield Palestine to the Zionists, the Arabs were concerned about the repeated, fervent Zionist attempts to displace them from their homeland by enlisting the support of the imperial powers. First Britain and then the United States came to the aid of the Zionists. On the face of it, they were interested in solving the Jewish question, but the heart of the matter was oil and their prime interest was to secure the area for themselves, if not directly then by proxy through dependencies and satellites. Israel was perceived by the Western powers as one such client state that would be dependent on their aid and therefore would necessarily have to serve their interests.

Zionist activity in Britain dates back to the early twentieth century. In 1900 the Fourth Zionist Congress was held in London because, as Herzl proclaimed, England was the only country in which Jews were not confronted with anti-Semitism. At the time, the British were disturbed over East European Jewish immigration into their country. Some British leaders were prepared to consider finding a home for the Jews in a part of the British colonies. Palestine was under Ottoman domination and thus was inaccessible for the time being. Indeed, the British offered Uganda to the Zionist leaders; at first, they accepted reluctantly and then, in 1907, retracted and announced that they would accept not substitute for Palestine.[1] There the matter rested until, during the First World War, Turkey joined the Central Powers. The Zionist leaders again turned to Britain in the hope that, when the Ottoman Empire was defeated, Britain would help them gain a foothold in Palestine.

During the First World War, three years — 1915, 1916 and 1917 — were of dramatic consequence for the future of Palestine, Zionism and the Middle East area. With the outbreak of the war, the Arabs had a golden opportunity to rid themselves of the Ottoman yoke. The Arabs'

55

attitude was of major concern to the Allies, and particularly Great Britain, which was quick to contact Sharif Hussein of Mecca, then considered the representative of the Arab nationalist movement. Britain's immediate aim was to secure Arab military support in the war against Turkey. To achieve this goal, formal negotiations were conducted between Sharif Hussein and Sir Henry McMahon, and the resulting agreement was known as the 'Hussein–McMahon correspondence'. The British promised their support for Arab independence if the Arabs, in return, would launch a revolt against the Ottoman Empire. Today all existing documents prove conclusively that Palestine was definitely included in the lands promised by Britain to Sharif Hussein. On the strength of this agreement, the Sharif gave the signal for the Arab Revolt which began on 5 June 1916.

In the spring of 1916, only six months after the British pledge to the Arabs, Britain signed with France (with Russian approval) the Sykes–Picot Agreement. According to this agreement, the Arab territories of the Fertile Crescent were divided into three areas of influence: Lebanon and Syria were to be under French control; Iraq and Transjordan to be under British control; and Palestine was to be placed under British mandate. This agreement clearly conflicted with the McMahon promises to Sharif Hussein. George Antonius has characterized the agreement in strong terms:

> The Sykes-Picot agreement is a shocking document. It is not only the product of greed at its worst, that is to say, of greed allied to suspicion and so leading to stupidity; it also stands out as a startling piece of double-dealing.[2]

However, the Sykes–Picot Agreement was not the last conspiracy the West had in store for the Arabs. In 1917 the British Foreign Minister issued a statement known as the Balfour Declaration which pledged that:

> His Majesty's government view with favour the establishment in Palestine of a national home for the Jewish people, and will use their best endeavours to facilitate the achievement of this object, it being clearly understood that nothing shall be done which may prejudice the civil and religious rights of the existing non-Jewish communities in Palestine, or the rights and political status enjoyed by Jews in any other country.

There was, in fact, no way in which a Jewish national home could be established without disturbing the existing non-Jewish communities in Palestine and prejudicing their civil rights. Whereas the Arab population was almost 700,000, there were no more than 56,000 Jews in Palestine in 1917, and Jewish ownership of land did not exceed 2 percent of the total land area of the country.

By the end of the First World War, Britain had acquired the mandate for Palestine. It is important to underline, however, that when the Balfour Declaration was issued in 1917, Britain was not yet in control of the country. When Britain assumed control, thousands of Jews immigrated to Palestine. While there were only 56,000 Jews in Palestine in 1917, their number had increased to 200,000 by 1932. During that period the Zionists were changing from mere missionaries to state-builders. With the League of Nations' approval of the mandate agreement in 1922, the Zionists acquired their first internationally binding pledge of support and their political claims to Palestine were greatly strengthened. In fact, the mandatory agreement was framed largely in the interest of the Jews. The Arabs, who constituted 90 percent of the population of Palestine, were referred to by the mandate agreement as 'the other section' of the population. Arab resentment simmered for many years, and finally broke out into open revolt in the mid-1930s. This uprising continued until the outbreak of the Second World War.

The London Conference and the 1939 White Paper

With the increase in Arab unrest and the growth of the Nazi threat in the Middle East, the British government became anxious to reach some understanding with the Palestine Arabs. In 1936 a Royal Commission was sent to Palestine on a fact-finding mission. The commission recommended the division of Palestine into a Jewish state, an Arab state, and a neutral area around Jerusalem to be kept under British administration. However, both the Arabs and the Jews rejected these proposals. In early 1939, with another eruption of Arab violence, the British government issued a policy statement on Palestine known as the MacDonald White Paper which pointed out that:

The Royal Commission and previous commissions of inquiry have drawn attention to the ambiguity of certain expressions in the mandate, such as the expression 'a national home for the Jewish people', and they have found in this ambiguity and the resulting uncertainty as to the objectives of policy a fundamental cause of unrest and hostility between Arabs and Jews.[3]

The White Paper recommended a reduction in Jewish immigration to Palestine and recognized some of the legitimate concerns of the indigenous Arab population. This represented a new British policy towards the Arab world. However, just as the British began to demonstrate growing sensitivity to Arab grievances and make some gestures of appeasement, so too was the awareness of Britain and other great powers of the strategic worth of the region's oil reserves being heightened.

As we saw earlier, Anglo-American rivalry over oil concessions in the Middle East must have had a particular influence on the new British policy towards the Arab world.[4] One immediate result was that Zionist leaders turned their attention to a new sponsor, the United States. This change was fortuitous, given the drastic change in political realities after the Second World War. Britain's power was greatly reduced and it was obliged to reduce its commitments in the Middle East accordingly. On the other hand, after the war, the Middle East was recognized as an area vital to American defense bases because of oil, communications and its strategic location. As a result, the United States has been drawn into every major crisis in the Middle East since 1945. Its first forays were on behalf of Zionism.

The United States and Israel

Zionist influence upon United States policy-makers dates from the First World War. With America's participation in the war, American political leaders began discussing the fate and aspirations of the Zionist movement in Palestine. In fact, President Woodrow Wilson not only sympathized with the purposes of Zionism, but actually referred to himself as a Zionist. The British government, which was aware of America's favorable attitude towards a Jewish National Home in Palestine, sought President Wilson's advice on, and approval of, the Balfour Declaration. After informal discussions and correspondence, President Wilson, acting through Colonel House, gave his approval of the declaration. It was subsequently published and received support from all Allied governments. On 31 August 1918 President Wilson expressed his personal sympathies with the Balfour Declaration in the following letter to Rabbi Stephen S. Wise:

I have watched with deep and sincere interest the progress of the Zionist movement in the United States and in the allied countries since the declaration by Mr. Blafour, on behalf of the English Government, of Great Britain's approval of the establishment of a national home for the Jewish people.[5]

On 2 March 1919 President Wilson showed his unequivocal support for the Zionist cause when he told a group of Jewish leaders:

As to representatives touching Palestine, I have before expressed my personal approval of the declaration of the British Government regarding Palestine. I am, moreover, persuaded that the allied nations, with the fullest encouragement of our government and people, are agreed that in Palestine there shall be laid the foundations of a Jewish commonwealth.[6]

In 1922 the Zionists succeeded in pressing for the 'Lodge-Fish Palestine Resolution', favoring the establishment of a National Home for the Jewish people in Palestine. This resolution was sponsored by both Houses of Congress. In his address to the Sixty-Seventh Congress on 30 June 1922, Senator Henry Cabot Lodge, the prime mover of the resolution, revealed his sympathy for the Jewish people and urged Congress to pass a resolution on the pretext that it 'neither threatens nor invades any rights of any other people'.[7] The resolution was then unanimously adopted by the Senate. After a heated debate, the final text of the Joint Resolution was passed on 11 September 1922, when the House accepted the Senate version. The resolution read:

Resolved by the Senate and House of Representatives of the United States of America in Congress assembled, that the United States of America favors the establishment in Palestine of a national home for the Jewish people, it being clearly understood that nothing shall be done which may prejudice the civil and religious rights of Christian and all other non-Jewish communities in Palestine, and that the holy places and religious buildings and sites in Palestine shall be adequately protected.[8]

The Joint Congressional Resolution of 1922 was regarded by the Zionists as only the beginning of American support. Support for the Zionist ideal grew in the 1920s and, in May 1932, culminated in the founding of the American Palestine Committee. The Committee declared that its function was:

> in keeping with the Lodge-Fish Joint Resolution of Congress, adopted
> in 1922, expressing approval of the United States for the establish-
> ment in Palestine of the national home for the Jewish people . . .
> [and] to foster the development of an informed public opinion in
> the United States among non-Jews concerning Zionist activities,
> purpose and achievements in Palestine . . . [the committee also
> regarded] the speedy realization of Zionism as an imperative neces-
> sity.[9]

The creation of the American Palestine Committee was immediately
followed by political support; the Committee soon enlisted 67 Senators
and 143 Representatives in its membership. Thus when the British
announced their 1939 White Paper, Congress reacted by demonstrating
its sympathy with the Zionist cause: 15 of the 25 members of the
House Foreign Affairs Committee and 28 Senators attacked the British
proposals. The American leaders called the defense of Jewish interests
in Palestine a 'moral obligation of the Unied States' and labelled the
new British policy a 'Treaty of Violation'.[10]

The British White Paper caused great suspicion among Zionist leaders
abut future British intentions towards Zionism. The Zionists quickly
saw the need to transfer their base of operations from Britain to the
United States:

> While they sought to eliminate the White Paper and if necessary
> Britain itself, and to overcome any threat from the Arabs, a third
> goal of the Zionists at this time was to win the goodwill and support
> of the United States, which they knew to be essential to the achieve-
> ment and maintenance of a Jewish state. This entailed, on the one
> hand, gaining the political support of the US Government, with a
> view especially to bringing pressure to bear on Britain, and, on the
> other, securing the interest, approval, and financial help of American
> Jewry.[11]

In an article entitled 'We Look Towards America', David Ben-Gurion,
the major advocate of this new strategy, wrote:

> I was certain that the center of decision was shifting from Europe to
> America . . . it was, therefore, essential to mobilize all the strength
> and influence of the Jewish people, and this — apart from Yishuv[12]

itself — was to be found primarily in America.

. . . only American Jewry could carry weight internationally if they were willing and able to mobilize their entire strength and influence. For America, right from the outbreak of war, played a most important role.

. . . this was the only Jewish Community that could wield an important — perhaps decisive — influence upon Britain's Palestine policy — if only it could mobilize all its strength in the struggle for the future of the Jewish people in 'its' homeland.[13]

Ben-Gurion acknowledged that achieving this aim would not be an easy task, because Jewish influence was declining in many areas of the world. Even though the Jewish community in Britain exerted tremendous pressure on the British government to retract their 1939 White Paper, nevertheless 'British Jewry's influence on its government's policy was practically nil.'[14] In the Soviet Union, the situation was not much different. Soviet Jewry, like other minorities, had been silenced since the Bolshevik revolution. In Europe, the Nazis gave the Jews little time to think of anything except their own survival.

The Jewish community in the United States (the world's largest), although it represented no more than some 3 percent of the total population, had achieved great political influence. However, the American Jews were not yet Zionists. As Moshe Shertok said at the time:

I think everyone agrees that America will have a decisive influence at the end of the war . . . and the question of our strength in America is a very real and important one. Weizmann . . . regards America as the only power, but this power is not utilized to further Jewish [Zionist] policy.

There are millions of active and well-organized Jews in America, and their position in life enables them to be most dynamic and influential. They live in the nerve-centers of the country and hold important positions in politics, trade, journalism, and theatre and the radio. They could influence public opinion, but their strength is not felt, since it is not harnessed and directed at the 'right target'.

The Zionist organization of America is at low ebb . . . the Jewish masses show deep feeling and a basic natural loyalty for our cause, but this feeling is not utilized and put to practical ends.[15]

Eliahu Golomb painted an even darker picture for the Zionist movement in the United States when he stated:

I must say that [the American Jew] today is neither willing nor able to fight for Zionism . . . a force can be crystallized from among American Jews for political action and practical aid for our cause. But so far it does not actually exist — it is only a potential force. To bring it into being much work needs to be done.[16]

The major task of the new strategy was to bring the Jewish community together, organize them to support Zionist aims, and direct their pressure towards the realization of a Jewish homeland in Palestine. As Moshe Shertok put it:

Perhaps our chances of success are not great, but the effort must be made. From now on this is not only a question of American Jewish influence being brought to bear on the American administration (a weighty issue in itself), but rather of American Jewry's influence being directed at the British government. Until now this has never been done.[17]

The year 1942 was a turning-point, as numerous Jewish and Zionist groups held meetings and organized protests and demonstrations against British policy in Palestine. Zionist successes culminated in the Biltmore Program of 1942.

Until the Second World War, the Zionists had deliberately avoided asking for a Jewish state and had requested simply a National Home. In May 1942 a major shift occurred. A conference was called at the Biltmore Convention in Baltimore, Maryland, by the American Emergency Committee for Zionist Affairs, with 600 delegates taking part. The conference sought to organize the Zionist Diaspora (those Jews living outside Palestine) into an effective body to implement the Biltmore Program:

The conference declares that the new world order that will follow victory cannot be established on foundations of peace, justice, and equality, unless the problem of Jewish homelessness is finally solved. The conference urges that the gates of Palestine be opened, that the Jewish Agency be vested with control of immigration into Palestine and with the necessary authority for upbuilding the country, including the development of its unoccupied and uncultivated lands; and that Palestine be established as a Jewish Commonwealth integrated in the structure of the new democratic world. Then and only then will the age-old wrong to the Jewish people be righted.[18]

This program went much further than the Balfour Declaration; the avowed Zionist objectives had now surfaced. By stepping up their demands, the Zionists expressed their growing conviction that a policy of moderation did not pay, that Britain could no longer be relied upon, and that it was expedient to seek the support of the United States.

Hand in hand with the Biltmore Program went intensive Zionist lobbying activity of the leading US politicians to support Zionist aspirations for:

Independence for Jews in Palestine and the building of an economy capable of supporting a vast influx of European Jews. Americans were the protagonists of this policy throughout the war years . . .

American Zionists were the most powerful political and economic body in the world movement and they intended to brook no compromises. The balance of power in the American Jewish community had become so altered that even the notables of the American Jewish committee, reluctantly, and with a due number of provisos, were ultimately swept along in support of the Jewish Commonwealth thesis which they traditionally opposed.[19]

Meanwhile, the systematic annihilation of millions of Jews by Nazi Germany led many people, both Jews and non-Jews, to conclude that the Jews ought to have a state of their own. Beginning in 1943, an intensive propaganda campaign was mounted by the American Zionists with a view to bringing American public opinion to support Zionism's aims. Finally, in January 1944, a resolution endorsing the Biltmore Program was introduced in both Houses of Congress, calling for concrete action on the part of the United States government to:

use its good offices and take appropriate measures to the end that the doors of Palestine shall be opened for free entry of Jews into that country, and that there shall be full opportunity for colonization so that the Jewish people may ultimately reconstitute Palestine as a free and democratic Jewish Commonwealth.[20]

This resolution was a landmark: American support for Jewish statehood was now a fact. However, Zionist efforts to obtain presidential endorsement of this resolution met with some resistance, and were greeted by the active disapproval of the Chief of Staff, General George C. Marshall. General Marshall objected that such a step might jeopardize the Allies' interests in the Arab world. Despite continued pressure on

President Roosevelt, the Zionists were largely unsuccessful in getting his full backing for their policies. However, during Roosevelt's administration, the precedent established under Woodrow Wilson, of direct White House interference in American policy towards the Middle East due to Zionst lobbying activities, continued. It was to reach new, higher levels under the Truman administration.

After the Second World War, high-ranking American officials were in complete accord with the Zionist drive for a new Jewish state. New congressional resolutions were introduced, and there were even some calls for unlimited Jewish immigration to Palestine. On 31 August 1945 President Truman appealed to the British Prime Minister Clement Attlee, asking for the immediate admission of 100,000 Jewish refugees to Palestine. Again in 1946, a congressional election year, President Truman issued a statement calling for the immediate admission of 100,000 Jews to Palestine. He also supported the Jewish Agency's proposal for the creation of a viable Jewish state with control of its own immigration and economic policies in an adequate area of Palestine. Two days later, the Republican presidential candidate and former Governor of New York, Thomas Dewey, declared that 'several hundreds of thousands of Jews should be admitted'. Thus Democratic incumbent and Republican opponent alike became further committed to the support of Zionism.

Palestine Before the United Nations

Meanwhile, the British position in Palestine was slowly becoming untenable. Subject to official American pressure, at odds with both the Zionists and the Arabs, and facing growing disorders in its mandated territory, the British government decided to take the problem before the United Nations. On 2 April 1947 Britain requested a special session of the General Assembly to consider the problem. The General Assembly, which met between 28 April and 15 May, set up a United Nations Special Committee on Palestine (UNSCOP). Between 26 May and 31 August 1947 UNSCOP held numerous public and private meetings, primarily in New York, Jerusalem, Beirut and Geneva. It studied reports, held hearings of persons and groups concerned with Palestine, made field trips to Palestine and some of the Arab states, and visited Jewish refuge camps in Europe.

Despite disagreement among the UNSCOP members, they unanimously agreed on eleven basic recommendations, the most important of which were:

(1) The Palestine mandate should be terminated as soon as possible and the country given independence.
(2) Provisions should be made for protecting the holy places.
(3) Steps should be taken to preserve the economic unity of Palestine.
(4) 'Any solution for Palestine . . . [should not] . . . be considered as a solution of the Jewish problem in general', and the General Assembly should help solve the Jewish refugee question by international action so as to 'alleviate . . . the intensity of the Palestinian problem'.

The UNSCOP report also contained two suggestions for the settlement of the Palestinian question. A majority plan, endorsed by Canada, Czechoslovakia, Guatemala, the Netherlands, Peru, Sweden and Uruguay, provided for Palestine to be partitioned into an Arab state, a Jewish state, and the international city of Jerusalem. A minority plan, proposed by India, Iran and Yugoslavia, advocated the establishment of a Palestinian Federation consisting of autonomous Arab and Jewish states. This plan suggested a three-year transitional period during which limited Jewish immigration would be allowed. The number of immigrants would be determined by a joint committee of Arabs, Jews and UN representatives, and also by the country's physical capacity.

The Arab states favored the minority plan because it satisfied their basic requirements, namely, a single independent state with an Arab majority and a limitation on Jewish immigration. The Zionists, albeit reluctantly, accepted the majority plan.

Both plans were debated by a special *Ad Hoc* Committee of the General Assembly at its fall session in 1947. On 29 November 1947 the General Assembly voted 33 to 13, with 10 abstentions, to recommend the partition of Palestine.

The United Nations decision in favor of partition was an example of high-level pressure politics in action. All the Zionist political machinery was primed to coerce the United Nations into creating a Zionist state in Palestine. US Secretary of State George C. Marshall expressed the United States' full support of the partition proposed on 17 September 1947 when he declared:

The United States gives a great weight not only to the recommendations, which have met with unanimous approval of the special committee, but also to those which have been approved by the majority of the committee.[21]

At the same time, President Truman put more pressure on his State Department than ever before to swing the United Nations vote for the partition of Palestine. In the words of the syndicated journalist, Drew Pearson:

Truman called Acting Secretary Lovett over to the White House on Wednesday and again on Friday, warning him he would demand a full explanation if nations which usually lined up with the United States failed to do so on Palestine. Truman had in mind such countries as Liberia, wholly dependent on the United States; Greece which would fall overnight without American aid; Haiti, which always follows Washington's lead; and Ethiopia, also indebted to the United States; . . . and all of these countries were stepping out of America's line on Palestine.[22]

Some of these countries were given instructions from their home governments to vote against partition. The Philippines delegation, whose government later gave in to United States pressure, delivered one of the most effective speeches against the partition resolution. In the speech, General Carlos Romula passionately defended the:

Inviolable, primordial rights of a people to determine the territorial integrity of their native land . . .
We cannot believe that the United Nations would sanction a solution to the problem of Palestine that would turn us back on the road to the dangerous principles of racial exclusiveness and to the archaic documents of theocratic governments . . .
The problem of displaced European Jews is susceptible to a solution other than through the establishment of an independent Jewish state in Palestine.[23]

A wave of telegrams was sent from 26 pro-Zionist Senators to the representatives of Haiti, Greece, Luxembourg, Argentina, Colombia, China, El Salvador, Ethiopia, Honduras, Mexico, Philippines and Paraguay, urging them to support the partition resolution. Most of these countries were either against partition, like the Philippines, or had no position on the issue. Affected by the official pressure from Washington, five countries changed their votes to 'yes' and seven changed their votes from 'no' to 'abstention'.[24]

The extreme pressure exerted on the State Department by American Zionists made Robert Lovett, the Under Secretary of State, declare that: 'Never in his life had he been subjected to as much pressure as he had in the three days beginning Thursday morning and ending Saturday night (during the final stage of the voting).'[25] Even President Truman, who gave Israel immediate *de facto* recognition, confessed in his *Memoirs*:

The facts were that not only were there pressure movements around the United Nations unlike anything that had been seen there before, but that the White House, too, was subjected to a constant barrage. I do not think I ever had as much pressure and propaganda aimed at the White House as I had in this instance. The persistence of a few of the extreme Zionist leaders, actuated by political motives and engaging in political threats, disturbed and annoyed me . . . [some Zionist leaders were] even suggesting that we pressure sovereign nations into favorable votes in the General Assembly.[26]

The forces favoring the establishment of the state of Israel were not denied. On 29 November 1947 the United Nations voted to recommend the partition of Palestine into an Arab and a 'Jewish' state. The Arabs refused to accept this plan. On 14 May 1948 the Zionists declared the establishment of the state of Israel. Within minutes, the United States gave Israel diplomatic recognition. Arab armies entered Palestine and the nagging conflict became a war.

There is no need to discuss the course of the conflict. It was ended by an armistice agreement in 1949, with Israel in possession of even more territory than had been allotted to it in the partition plan. Thus the third stage of the Zionist movement, the invasion stage, was achieved. A few months later, with American help, Israel was admitted to the United Nations.

No final words foresaw the future implications of this tragedy as well as those of Sir Zafrullah Khan, Foreign Minister of Pakistan, during the United Nations debate:

Those who have no access to what is going on behind the scenes have known enough from the press to have fear in their hearts not only on this question — because this is one individual question — but that the deliberations on crucial questions of this great body, on which the hopes for the future are centered, will not be left free . . .

Remember, nations of the West, that you may need friends tomorrow, that you may need allies in the Middle East. I beg of you not to ruin and blast your credit in those lands.[27]

Notes

1. Herzl, who was the main advocate of accepting Uganda as a temporary measure, died in 1904. He was succeeded by David Wolffsohn, who was ousted in 1911. From that time on, the World Zionist Movement was dominated by Dr Chaim Weizmann, although the new President was Otto Warburg, a leading scientist from Hamburg.

2. George Antonius, *The Arab Awakening*, 5th edn (Capricorn Books, New York, 1965), p. 248.

3. Robert John and Sami Hadawi, *The Palestine Diary*, vol. 1 (Palestine Research Center, Beirut, 1970), p. 315.

4. As Professor Hamed Rabie observed, 'Oil represents a core factor in the formation of major events concerning the Middle East area.' *The Oil Weapon and the Arab-Israeli Struggle* (Arab Institute for Research and Publishing, Beirut, 1974), p. 86. Thus it is reasonable to suggest that the change in British policy in Palestine was motivated by the increased British concern over the new Arab oil wealth.

5. Ray S. Baker, *Woodrow Wilson, Life and Letters*, vol. 8 (Doubleday, Page & Co., Garden City, N.Y., 1927), pp. 372ff. It should be noted that President Wilson's declaration of American sympathy with Zionism was delayed not because of any hesitancy on the part of the American government, but owing to the fact that the United States and Turkey were not formally at war.

6. Carl Hermann Voss, *The Palestine Problem Today: Israel and its Neighbors* (Beacon Press, Boston, 1953), p. 11.

7. US Congress, House Committee on Foreign Affairs, *The Jewish National Home in Palestine*, address by Senatory Henry Cabot Lodge, 30 June 1922, 78th Cong., 2nd sess., referred to in Hearings, 8, 9, 15, 16, Feb. 1944, p. 375.

8. Quoted in Frank E. Manuel, *The Realities of American Palestine Relations* (Public Affairs Press, Washington, D.C., 1959), p. 282.

9. Quoted in Fink, *op. cit.*, p. 59.

10. Manuel, *American Palestine Relations*, pp. 307-8.

11. Allen, *op cit.*, pp. 257-8.

12. The Jewish population of Palestine.

13. David Ben-Gurion, 'We Look Towards America' in Walid Khalidi (ed.), *From Heaven to Conquest* (Institute for Palestine Studies, Beirut, 1971), pp. 481-8.

14. Ibid., p. 483.

15. Ibid., p. 484.

16. Ibid.

17. Ibid.

18. Biltmore Declaration, 11 May 1942, quoted in Fred J. Khouri, *The Arab-Israeli Dilemma* (Syracuse University Press, Syracuse, 1974), pp 29-30.

19. Manuel, *American Palestine Relations*, p. 310.

20. *The Jewish National Home in Palestine*, referred to in Hearings, 8, 9, 15, 16, Feb. 1944, p. 391.

21. Joseph Dunner, *The Republic of Israel* (McGraw-Hill, New York, 1948), p. 78.

22. John and Hadawi, *Palestine Diary*, vol. 2, p. 262.

23. Alfred M. Lilienthal, *There Goes the Middle East* (Devin-Adair Co., New York, 1957), pp. 4–5.

24. Ibid., p. 7.

25. Alan R. Taylor, *Prelude to Israel: An Analysis of Zionist Diplomacy, 1897-1947*(The Philosophical Library, Washington, DC, 1959), p. 103.

26. Harry S. Truman, *Memoirs, vol. 2: Years of Trial and Hope* (Doubleday, Garden City, N.Y., 1955), p. 158.

27. Quoted in John and Hadawi, *Palestine Diary*, vol. 2, p. 263.

7 ISRAEL AND WESTERN STRATEGY IN THE MIDDLE EAST

Even though they have experienced bitter periods of colonialism, suppression and exploitation at the hands of the West, after independence the Arabs sought to reaffirm relations with Western powers on friendly terms. Arab–American relations, in particular, have a long and colorful history. But until the end of the Second World War, American interests in the area were limited largely to cultural and educational affairs, as their concern for oil was slow to take shape. It was not until the post-war era that the United States became committed to a political and military presence in the region. Arab–American relations changed as both Arab and American policies came to revolve around the Palestinian question. Every major United States policy decision concerning the Arab world has been influenced to some degree by the American position on Palestine.

Indeed, the establishment of Israel has alienated most of the Arab world from the West. The Arabs see Israel as a Western creation, and blame mainly the United States and, to a lesser extent, Britain for assisting the Zionist movement to establish the Jewish state:

> The image that emerges is of an Israel that is at once an American outpost on a distant frontier and a staunch, *independent* ally; an Israel, to use the *Guardian*'s words, which is 'America's sheriff in the Middle East [manning] the frontiers of the free world against the encroachment of Soviet imperialism'. In this framework the United States would presumably be committed to Israel's safety as France had been to French Algeria's – America as Israel's 'Metropole'.[1]

America's adoption of Israel as its flagship in the Middle East represents the principal background factor to relations between the Arabs and the West on the one hand, and Israel and the West on the other.

The roots of Israeli foreign policy extend back to the beginning of the modern Zionist movement. In particular, reliance upon diplomatic alignment, and force when necessary, to deal with Middle Eastern opposition to Israel is the fundamental facet of Israeli foreign policy, whose origins go back to before the First World War. Herzl and Weizmann stressed the strategic significance of Palestine, which guards the

70

eastern flank of the Suez Canal and which could serve as a source of regional influence for its sponsor. Like the Zionist movement before it, the state of Israel has depended on great-power sponsorship. Israeli leaders contended that Israel's existence and prosperity would benefit its patron, as Israel could perform regional diplomatic and military tasks which its protector would find inconvenient, and possibly dangerous, to undertake directly.

To be more specific, many Israeli officials have suggested that the United States, through a 'strong Israel', maintains a certain measure of control over political and territorial developments in the Middle East and at the same time blocks the growth of Soviet influence in the region. In General Moyshe Dayan's view, as reported in the Israeli newspaper *Haaretz*, the basis of the Israeli–American relationship was threefold:

(1) Israel could do its own fighting; all it needed from the United States was armaments, not soldiers.

(2) A militarily strong Israel would make the Arabs despair of a military solution, thus preserving the peace and preventing the possibility of a great-power confrontation in the Middle East.

(3) Israel could best serve American interests in the Middle East — mainly the oil flow — by protecting the conservative Arab regimes in the [Arabian] Gulf from the Arab radicals, such as Egypt and Syria.

As the *Haaretz* article put it, 'We explained that the Israeli army, with its real and not just relative power, presents a first line of defense for American interests in the Mediterranean area.'[2]

In the United States this theory of the indispensable nature of Israel to Western, particularly United States, interests in the Middle East is also advanced by some American officials. In the words of Senator Henry M. Jackson of the state of Washington:

The United States actively participated in helping to create the State of Israel, and since its founding the people of this young nation have won the admiration of the great majority of Americans by the valor they have demonstrated in standing firm before their hostile neighbors. Unlike some countries of the Middle East, Israel is a stable democracy, and a profoundly egalitarian and spirited one. These qualities, too, inspire the respect of many Americans, who feel something like a sense of personal involvement in the destiny of Israel.

Today, Israel is serving as the front line of Western defense in the Middle East.[3]

(Emphasis added)

Thus the first item of business on the Western agenda, after the establishment of Israel, was to introduce measures to guarantee the continuous existence of this 'young, spirited democratic' state. In April 1949, with the signing of the North Atlantic Treaty Organization pact (NATO), which later came to include Turkey and Greece, the Western powers began to organize a political-military bloc in the Middle East. Israel, consequently, was regarded as the cornerstone of security and hegemony in the area. With the Tripartite Declaration of 25 May 1950, the United States jointed with Britain and France to ensure Western gains in the area and preserve the *status quo* by ensuring Israel's security. The three powers declared their support of the armistice agreements of 1949 between the Arab states and Israel 'in order to safeguard their internal security and their legitimate defense' and to enable them 'to fulfil the task assigned to them [i.e., to the Arabs and Israel] in the defense of the entire region'.[4]

Pressure was brought to bear on the major Arab states to participate in successive military plans, ranging from the Middle East Defense Organization (MEDO) in 1951 to the Eisenhower Doctrine in 1957. The West's strategy was designed to include the Arab states within the global disposition of the West, under the guise of countering Soviet threats. These plans foundered on the Arabs' strong anti-Western political sentiments which developed with the creation of Israel. From the perspective of the Arab states themselves, the West's policies were looked on primarily as devices to maintain Western influence over the area and to hinder efforts to unify the Arab countries. Moreover, throughout the Arab world, there was — and continues to be — a deep belief that the real enemy is Israel and the West, not the Soviet Union or communism. As Ali Abdul-Nawar, who replaced Brigadier Glubb in 1955 as Commander of the Arab Legion of Jordan, stated:

We do not know that it is more vital to defend ourselves against Russia than the West. We have only heard about communism . . . We have had no experience of it, as we have had of Western domination.[5]

Aside from Israel, the key to successful American policy in the Arab world was Egypt. President Nasser was seen as a major obstacle to any possibility of improvement in Arab-United States relations, and Egypt's refusal to participate in any defense scheme posed a major problem for United States policy in the Middle East. The watershed came in 1955, when President Nasser announced the Soviet-Egyptian arms agreement. This development, which broke the 'Western arms monopoly', was regarded at the time as the great turning-point in the Middle East — the end of one era and the beginning of another. The deal was officially finalized on 27 September 1955. President Nasser explained this agreement in April 1955:

> We asked the Soviet Union for help. For years we had asked everyone else. We turned to the Soviet bloc knowing that Western powers had no intention of dealing with us as independent equals. Help came from the Soviet bloc almost as quickly as the ink dried on our request.[6]

The State Department's immediate reaction was to show Nasser that he still needed the United States. In June 1956 Washington withdrew its offer to finance the Aswan High Dam. Nasser retaliated by nationalizing the Suez Canal Company. The deterioration in Anglo-American relations with the Arab world reached its nadir with the Suez crisis in 1956. Israel proved to the West its viability as a tool of Western interests and was left as their only reliable ally in the Middle East.

Although the United States was consulted throughout the crisis, President Eisenhower was not informed of military action before the actual attack. Both Prime Minister Eden of Britain and Prime Minister Mollet of France feared that Washington would not approve of their actions and that their planned schedule for military operations would be thwarted. The Israeli-Anglo-French attack against Egypt was condemned by the United States. On 29 October 1956, on the evening of the Israeli invasion of Egypt, President Eisenhower immediately called a special meeting of his top advisers at the White House. He later described what took place at the meeting:

> Some . . . saw the Israeli attack as a probing action, while others believed it would be a rapid move which would take the Israeli forces to Suez within three days at the most, and this would be the whole affair. Foster disagreed with them both.
>
> Foster Dulles said the situation was more serious; if the Canal were disrupted and pipelines broken, he thought the French and British would intervene. He suspected that they already planned to

do so.

Some at the meeting speculated that the British and French might be counting on the hope that when the chips were down, the United States would have to go along with them, however much they disapproved. But we did not consider that course.[7]

To Professor Hisham B. Sharabi, the American position during the whole episode was a critical one:

The US, once committed to action through the United Nations, had to pursue this line of policy to its logical conclusion. Dulles had little choice. The Hungarian uprising and strong US opposition to Soviet intervention there made it impossible for the United States to suport or even take a neutral position toward the tripartite attack on Egypt . . .

It must be remembered that US policy had supported decolonization in Africa and Asia and could not now risk being identified with British and French imperialism. From the point of view of the former colonies and dependent countries, the invasion of Egypt represented a resurgence of European imperialism, which confirmed the theories of *neo*-colonialism and *neo*-imperialism. Facing the Russian intervention in Hungary, and to preserve its position *vis-à-vis* the Afro-Asian bloc, the United States had no choice but to back the call for an immediate cease-fire; to condemn aggression; and to demand the unconditional withdrawal of the British, French and Israeli troops.[8]

But the strong United States opposition to the British–French–Israeli aggression against Egypt was primarily tactical. It should not be allowed to obscure the shared objective: to isolate and if necessary eliminate Nasser, and consequently Egypt, as a political force in the Middle East (an objective which American policy achieved later during the Sadat era with the Camp David Agreement). The United States correctly saw that direct military action would enhance Nasser's prestige among the Arab masses and increase political obstacles to Western strategy in the Arab world.

Soon after the crisis, America's real objectives came to light in the Eisenhower Doctrine, which constitutes a milestone in United States involvement in the Arab world. As Professor Sharabi put it:

The Middle East had just emerged from a war in which not Communism but Western imperialism and Zionism had shattered the peace of the Middle East and exposed the world to the threat of global conflagration. The Arab countries, even the most moderate and pro-Western among them, found it difficult to subscribe to the new doctrine, which overlooked the immediate threat and addressed itself to a danger by which no Arab state felt directly threatened.[9]

The Eisenhower Doctrine stressed the following points:

(1) A commitment on the part of the United States to support the independence and sovereignty of *all* Middle Eastern states threatened by communist control.

(2) The use of American armed forces — authorized by Congress — to protect any nation or group of nations in the Middle East *requesting* such aid against overt communist aggression.

(3) Securing the co-operation of Middle East states in the program to assist them in developing their economic strength in order to maintain their national independence.

In the words of Quincy Wright, this doctrine only served to convince the Arab countries, 'which had just experienced aggression, not from international communism but from Israel, Great Britain and France, that the United States had changed its policy which had opposed the latter aggression'.[10]

From that time onward, American policy in the Middle East during the Eisenhower administration became a continued effort to implement the Eisenhower Doctrine and to convince the Arab world that the United States was indispensable for the security of the Middle East. However, of the Arab states, only Lebanon formally adhered to the Eisenhower Doctrine. Without going into lengthy detail, the major events which took place in the Middle East from 1957 until the end of the Eisenhower administration in 1961 were either directly or indirectly linked to the Eisenhower Doctrine. The Syrian crisis in August 1957, the Jordan upheaval in April of the same year, the Lebanon civil war in 1958 — all were incidents which the United States seized upon as excuses to employ the Eisenhower Doctrine.[11] However, it was clear that the United States was unable to enforce its wishes upon the Middle East, and the net result of America's intervention, therefore, was to provide more grist for the mill of Soviet and anti-Western Arab forces. Senator John F. Kennedy described US and Western policy in the Middle East in the following manner:

The Middle East today is a monument to Western misunderstanding. During the last eight years the West has ignominiously presided over the liquidation of its power in the whole region, while the USSR has gained important footholds. American policy has wavered and wobbled as much, if not more, than any other Western country.

The situation was not improved by the dramatic announcement of an Eisenhower Doctrine to be defended against external aggression when, in fact, the indigenous Arab nationalist revolution and internal Communist subversion were the crucial factors . . .

Our mistakes in the Middle East, it seems to me, were primarily mistakes of attitude. We tended to deal with this area almost exclusively in the context of the East–West struggle – in terms of our own battle against international Communism. Their own issues of nationalism, of economic development, and local political hostilities were dismissed by our policy-makers as being of secondary importance . . .

The Arabs knew that their lands had never been occupied by Soviet troops – but that they had been occupied by Western troops – and they were not ready to submerge either their nationalism or their neutrality in an alliance with the Western nations . . .

In short, from here on out, the question is not whether we should accept the neutralist tendencies of the Arabs, but how we can work with them. The question is not whether we should recognize the force of Arab nationalism, but how we can help to channel it along constructive lines.

The mistaken attitudes of the past – our previous misconceptions and psychological barriers – must all be junked – for the sake of the Arab and for our own sake as well . . . We must talk in terms that go beyond the vocabulary of Cold War – terms that translate themselves into tangible values and self-interest for the Arabs as well as ourselves.[12]

During his short-lived administration, President Kennedy worked to improve US–Arab relations. However, Kennedy's Arab policy did not have time to mature. The honeymoon period enjoyed during the Kennedy administration ended soon after his death in 1963. Although the Johnson administration's involvement in Vietnam was to put every other issue, including the Middle East, into the background:

Johnson's attitude toward the Arab world, as contrasted to Kennedy's, seems to have been greatly influenced by his 'provincial' view of politics . . . On the whole, [he] as [John S.] Badeau put it, had 'little patience toward Arab affairs'.[13]

By the end of 1966, it was evident that United States policy was changing from one of limited friendship and normal relations to one of hostility. Strains in relations between the two sides reached a peak during the 1967 crisis. Nasser was considered by both the United States and Israel as the major obstacle to Western hegemony over the Arab world. Ever since the Suez crisis, Nasser had been leaning increasingly towards Moscow. In addition to his pan-Arab aspirations, Nasser was also attempting to establish himself as the main political figure in the emerging Third World. In an exclusive article in *Penthouse* magazine, Anthony Pearson asserted that by 1965 officials in the American government and Israel were in agreement that Nasser would definitely have to go; the issue was how to do it. As Pearson observed:

It was impossible to overthrow [Nasser] from within Egypt by any sort of coup . . . Nasser's constant and increasing threats against Israel and his assurance to his people that the Israel menace would be defeated suggested to the CIA that an Egyptian loss of face could be achieved by calling Nasser's bluff – by actually making him confront Israel . . .

At a series of secret meetings in Tel Aviv and Washington [between American and Israeli officials, including the CIA] it was decided to promote a contained war between Israel and Egypt – a war that would not affect the territorial lines between Israel and Syria and Jordan.[14]

The June 1967 war went as planned. Israel launched its pre-emptive attack against Egypt, and the Egyptian Air Force was destroyed in the early morning of the first combat day. Israel was victorious within a few hours. On the fourth day of fighting, Jerusalem fell and Jordan was defeated. Now only Syria remained. Even though the United States' major objective had already been achieved – Nasser's defeat – Israel, given its expanionist nature, could not resist the opportunity to attack Syria, with full United States support. But the Israeli plan was not consistent with the joint plan agreed upon with the United States, which called for a contained war. Instead, the Israelis followed their own determined plan for reshaping the Middle East in Israel's own interests. But, first, as Pearson has disclosed, one obstacle – the USS *Liberty* – had to be removed unless the United States should discover

Israel's violation of the original plan:

> On 8 June Israel was still three days away from her final objectives. It was possible that drastic measures might have to be taken at any time. The Israeli leaders were afraid that the continued presence of the *Liberty* off Sinai, monitoring their activities for both the US government and the United Nations, might wreck their plans. If the ship was sunk with all hands, the attack would be blamed on the Egyptians, or perhaps on Russian fighters from a Soviet fleet carrier. It would also serve the purpose of involving the Americans directly and committing them totally to Israel's side. It was a daring plan — a vicious plan — but certainly well-coordinated and well-executed. It seems surprising that it failed. Everything seemed stacked against the *Liberty*.[15]

Some Arab writers have also aired suspicions of possible Anglo–American collaboration with Israel during the June 1967 war against Egypt, Syria and Jordan. Professor Hisham B. Sharabi observed:

> The Arab States accuse the United States and Britain of collusion with Israel. Both Governments emphatically deny this. Just as the 'Suez Canal Conspiracy' came to light gradually after 1956, so also will the happenings before and during the June 1967 war become known in due time. But in the meantime, military experts indicate that Israel could not possibly have carried out a round-the-clock attack on twenty-five airfields in the United Arab Republic, covering distances of some 2,500 miles, and in addition attacked targets in far-off Iraq and Latakia and covered the Syrian and Jordan fronts all at once without receiving additional aircraft and having closer bases on land and aircraft carriers at sea from which to operate. Even if this were possible, the experts point out that this vast air operation could not have remained undetected by the US Sixth Fleet and the British warships in the Mediterranean . . . The fact that neither the US Sixth Fleet nor the British ships warned the UAR of the impending attack means one of two things: either the radar instruments of both fleets are obsolete, or that there was enough collusion to allow the surprise attack to succeed. To any fair-minded person, the latter would be more convincing. The very fact that the United States warned the UAR not to fire the first shot and later supported the Israeli position is in itself collusion.[16]

In the aftermath of the 1967 war, the question that remains to be answered is: How much has Israel contributed to the preservation of Western, particularly American, interests in the Middle East? In order to answer this question, we must first identify the United States' four fundamental policy objectives in this area:

(1) First, and most important, to prevent the area from falling under the control or influence of the Soviet Union, or any other major powers hostile to America.

(2) To protect the maintenance of air and sea transit rights for the United States and its allies. Of utmost importance is the freedom of movement of the American Sixth Fleet in the Mediterranean. The Suez Canal is also of vital importance to commercial traffic, particularly the transit of industrial goods from Western markets to the area and of oil from the Middle East, particularly the Arabian Gulf, to Western markets. The strategic location of the canal also makes it of great military value to the Western world.

(3) To maintain the flow of oil from Middle Eastern oilfields to Western markets.

(4) To preserve peace and security in the area. This policy relates, to a great extent, to the state of Israel. As Robert E. Hunter put it: 'The United States has a very special relationship with this country [Israel] that has come about because of an historical connection. Quite simply, the United States had a Palestine policy long before it had a Middle Eastern policy.'[17]

To return to our question, did the United States succeed in achieving these major objectives? In the words of William R. Polk:

It was obvious that the United States had failed to achieve or maintain these interests. The June 1967 war altered the whole face of the Middle East. At that time, the overriding American worldwide objective was avoidance of direct confrontation with the Soviet Union. This was achieved by tacit agreement of the two superpowers and by the speed and course of events in the Middle East. While the Middle East did not go under the control of the Soviet Union, the ruptured diplomatic relations between the United States and the Arab countries, the increased polarization of the area along Soviet–Arab and Israeli–Western lines, and the massive Soviet entry into the Middle East, particularly in Syria, Iraq and Egypt, certainly put this objective in jeopardy.[18]

The second American objective was largely lost. The Suez Canal was closed at vast cost to the United States and other countries. Until October 1973 the third factor, the flow of oil, proved largely unaffected by the level of crisis.

Even before the eruption of the fourth Arab-Israeli war in October 1973 and the beginnings of the Arab use of oil power, many Americans were urging a review of United States policy in the area. However, as John H. Davis argued:

[O] ne of the primary obstacles for most Americans in understanding the nature and history of problems in the Middle East is the difficulty and risk of raising reservations about policies of the Israeli Government or our relationships with Israel. In these areas, rational discussion most often leads to emotional rhetoric and the questioner being considered pro-Arab, anti-Israel, or worse. In effect, there has never been in our country any broad, significant public debate on the basic issues of Arab-Israeli relations and how our country might contribute most productively to improving them. About the only national attention to these subjects comes periodically in Congress and every four years when Presidential candidates and numerous Senators and Congressmen engage in a competition to see who can promise the highest level of American support to Israel.[19]

Israel was and continues to be an advanced military post, a 'garrison state' for the West in the heart of the Arab lands. Its main purpose for the West is to suppress the awakening of Arab political aspirations and to prevent the Arabs from controlling their own resources. Events, however, have proved that Israel too has its own Zionist designs for the area, and is equally capable of exploiting its privileged position in the West to serve its narrow interests of expansion over neighboring territories, even when these interests are in conflict with Western interests and objectives. This exploitation of the West by Israel has gone so far as to destroy the West's relationship with the Arab world. The West's prestige in the Arab world decreases in proportion to Israel's success in expanding its influence over neighboring Arab land. The main loser in this game has been the United States, which stands to lose not only in influence but also economically and strategically. On the other hand, the Arabs have consistently resorted to employing their economic leverage based on oil to seek world understanding and accommodation, and they will continue to do so. These issues are constantly on the agenda of the Arab League.

Notes

1. Hisham Bashir Sharabi, *Palestine and Israel: The Lethal Dilemma* (Pegasus, New York, 1969), p. 33.

2. *The Link*, vol. 7, no. 1 (Jan./Feb. 1974), p. 2.

3. Henry M. Jackson, *The Middle East and American Security Policy*, Report to the Committee on Armed Services, US Congress, Senate, 91st Cong., 2nd sess., Dec. 1970 (US Government Printing Office, Washington, D.C., 1970), p. 2.

4. US, Congress, Senate, Committee on Foreign Relations, *A Select Chronology and Background Documents Relating to the Middle East*, 89th Cong. (US Government Printing Office, Washington, D.C., 1967), p. 64.

5. Emil Lengyel, *Egypt's Role in World Affairs* (Public Affairs Press, Washington, D.C., 1957), p. 42.

6. David Duncan, 'Nasser, Frank and Startling Views', *Life* (16 April 1956), p. 34.

7. Dwight D. Eisenhower, *Waging Peace: 1956-1961*, quoted in Sharabi, *Palestine and Israel*, p. 59.

8. Sharabi, *Palestine and Israel*, pp. 60-1.

9. Ibid., pp. 64-5.

10. Quincy Wright, 'Legal Aspects of the Middle East Situation', *Law and Contemporary Problems*, no. 29 (Winter 1968). Quoted in Sharabi, *Palestine and Israel*, p. 65.

11. In his memoirs, President Eisenhower acknowledged that plans were drawn up by American strategists in Syria and Jordan to create an environment in which the Eisenhower Doctrine could be employed. Syria was to have been declared under communist control; a plot within the Syrian Army would be staged in which exiled Syrian leaders would give their support; an appeal would be made for help from sister Arab states, and contingents of the Iraqi and Jordanian armies would invade Syria from the south and east − and, if necessary, Lebanese and Turkish help would also be made available. The role of the US was to remain limited to supplying arms and money, and to undertaking covert operations inside Syria. See Sharabi, *Palestine and Israel*, p. 67, and Joe Stork, *Middle East Oil and the Energy Crisis* (Monthly Review Press, New York, 1975), pp. 80-1.

Sharabi has argued that there was no basis whatever for American intervention in Lebanon. Even though President Chamoun appealed to the United States for help, such action could not be taken in the presumption of the Eisenhower Doctrine. As Sharabi suggested:

In requesting US intervention, President Chamoun invoked the Eisenhower Doctrine. The Doctrine had specified that the United States was willing to come to the aid of countries threatened by international Communism or 'Communist subversion' − not by internal political opposition. The 'rebels' in Lebanon were obviously not Communist . . .

The precipitate landing of American Marines in Lebanon was thus occasioned in reality, not by what was going on in Lebanon but by the military coup in Iraq. It was not, as Murphy put it, because Lebanon 'had accepted the Eisenhower Doctrine and hence was in a position publicly and internationally to invoke it' that the United States intervened. (pp. 68-9).

12. John F. Kennedy, *The Strategy of Power* (Harper & Row, New York, 1960), pp. 106-8.

13. Sharabi, *Palestine and Israel*, p. 78.

14. Anthony Pearson, 'The Attack on the USS *Liberty*: Mayday! Mayday!', *Penthouse* (May 1976), p. 143.

15. Pearson argued that the presence of the USS *Liberty* was an unknown factor in the joint American–Israeli plan. Thus, the Israelis looked at its presence as a liability, a factor which hindered their special plans. Ibid., p. 145.

16. Sharabi, *Palestine and Israel*, p. 314.

17. Robert E. Hunter, 'American Policy in the Middle East', *Asian Affairs*, no. 5 (1968–9), pp. 265–6.

18. William R. Polk, *The United States and the Arab World*, 3rd edn (Harvard University Press, Cambridge, Mass., 1975), p. 415.

19. John H. Davis, 'America's Stake in the Middle East', *The Link*, vol. 9 (Summer 1976), p. 2.

8 THE ARAB LEAGUE AND ARAB OIL POLICY

The creation of Israel in Arab Palestine in 1948 usurped what is to the Arabs an integral part of the Arab world and encouraged Arab statesmen to band together in the face of a common threat. The Arab League, established in 1945, became a more compelling Arab political force in 1948. The presence of an expansionist Western client state in their midst was, and has remained, a matter of grave concern to Palestinians and other Arabs, and has markedly influenced the area's economic and political life.

Determined to face the challenge, five members of the Arab League (Egypt, Lebanon, Saudi Arabia, Syria and Yemen) signed a collective security pact known as the Mutual Defense and Economic Co-operation Treaty (MDECT) on 17 June 1950. Other Arab countries, including Algeria, Tunisia, Libya and Sudan, joined the treaty in 1969. MDECT advocated close collaboration among its members in economic and financial matters. The importance of co-ordinating the exploitation of natural and industrial resources, especially oil, was emphasized in paragraph (d) of Article 1 of the treaty, which authorized:

> The submission of recommendations for the exploitation of the natural, industrial, agricultural and other resources of the contracting countries and the co-ordination of such exploitation in the service of the military efforts and mutual defense.[1]

Thus the treaty was primarily intended to meet external threats; economic co-operation among the member states was originally sought as a by-product of joint defense. This provision was to prove significant for oil, the most important natural resource in the Arab world, as subsequent events have shown.

Ever since Middle Eastern oil became 'a decisive if not vital commodity for the Western powers and hence of great international political value', Arab nationalists have striven hard to politicize Arab oil and bring it under the League's political influence. As early as the mid-1940s, a wave of demands called for oil to be used as a weapon against Western countries which supported Zionist aspirations for a state in Palestine.

When the Arab League was established in 1945, it considered the creation of a political body composed of Arab League members which could achieve harmony and co-operation among the member states' oil policies. The League's Charter was signed on 22 March 1945 by the heads of state of the then independent Arab countries: Egypt, Lebanon, Saudi Arabia, Syria, Transjordan and Yemen. A Palestinian delegation participated with observer status, since Palestine was then under British mandate. Article 2 defined the objective of the newly-founded organization:

> The League has as its purpose the strengthening of the relations between the member states; the co-ordination of their policies in order to achieve co-operation between them and to safeguard their independence and sovereignty; and a general concern with the affairs and interests of the Arab countries.

The objectives of the League were mainly 'defensive', but the League also advocated the close co-operation of member states in economic, financial, social and health affairs. Co-ordination of the Arab League members' oil policies was one of the primary economic goals. As Western support for the Zionists' aims increased, the Arab League began to change its political stance. At its Bludan meeting in Syria in June 1946, the League passed a set of resolutions which called for Arab oil to be used as a weapon. Thus, in 1947, concerned by the implied Arab threat, the US State Department advised against President Truman's support of the partition (of Palestine) plan.

However, because of the Arabs' total dependence on oil as their major source of income and on foreign companies to run the oil operations, the Bludan resolutions were never given serious consideration. In fact, some Arab countries still believed that political considerations should not be involved in commercial oil operations.

Immediately after the creation of Israel, the pipelines which carried oil from Iraq to Haifa for refining or trans-shipment to Western Europe were cut off. To this day these pipelines remain unused. When Aramco built the Trans-Arabian Pipeline (Tapline) from its fields on the Arabian Gulf Coast to the Mediterranean, it was forced to avoid Israeli territory even though it would have meant a shorter route. The Arab countries also put pressure on the international oil companies not to associate with the Israeli oil industry and, as a result, Israel was forced to build its own refining operations and to seek oil mainly from non-Arab countries like Iran and Venezuela.

However, the Arab campaign in 1948 was unsuccessful in pressuring

Western countries and major oil companies to disassociate themselves from Israel. In spite of all Arab threats, the Western nations voted to divide Palestine; and after Israel had occupied areas larger than those allocated to it by the United Nations, no Western country raised a finger to object. The Arab campaign failed:

> Whether a more concerted campaign could have been mounted is highly doubtful, given the low level of technical competence and political collaboration in and between the Arab states, and the limited dependence of the industrialized countries of that time on Arab oil. In any case . . . the threat remained dim and distant.[2]

The League continued to complain that its member states had no say in the fundamental problems facing the oil industry. They had no control over production, prices or marketing. Moreover, all the tankers, refineries and markets were owned by the major companies and, under these circumstances, only the companies were able to dispose of crude oil and refinery products. As a result, the Arab League concluded that the existing concessions not only allowed too great a latitude to the companies, but also prevented adequate supervision of their activities by the host governments.

Therefore, the League decided to adopt some fundamental decisions affecting the organization and political use of oil. At its August-September meeting in 1951, the Political Committee of the League decided to establish an Arab Oil Exports Committee (AOEC), which was set up the same year and became the League's main body for the formulation of oil policies. As Professor Zuhair Mikdashi stated:

> The reason behind the establishment of the committee [AOEC] was primarily political, namely the national security of Arab countries. This meant the protection of the political independence and territorial integrity of Arab states. It was hoped it would be achieved at that time by instituting a general economic boycott, including a boycott of oil supplies to Israel with a view to cutting the latter's military capacity for expansion into neighboring Arab countries.[3]

In subsequent years, the AOEC adopted many recommendations, some of the most important of which are the following:

(1) the establishment of Arab national companies and the support of those already in existence, including the creation of an Arab oil-

tanker company and the study of the economic feasibility of an Arab company for pipelines;

(2) the setting up of joint Arab refineries;

(3) the establishment of Arab petrochemical and other oil industries;

(4) the marketing of surplus oil products in Arab countries and the increase of oil trade and exchange among the Arab countries themselves;

(5) the unification of Arab oil terminology;

(6) the exchange of information and expertise in the oil sector.[4]

Nevertheless, one of the primary tasks which the AOEC took up was the institution of a full Arab boycott against Israel. Because Israel did not comply with the 1952 UN resolution giving the Palestinian Arab refugees the right to return to their homeland, the League's Political Committee adopted a resolution proposed by the AOEC which stated:

In view of the legitimate right of self-defense, considering the necessity of limiting Israel's power of aggression, noting Israel's persistence in violating the resolutions of the UN, and the latter's action in overlooking Israel's violations of the resolutions, considering that all these reasons make it incumbent upon the member-states of the Arab League to maintain complete solidarity in tightening the grip of boycott on Israel. And after being notified of the conditions and restrictions imposed by the League's member states on firms, tankers and ships to ensure that petroleum does not reach Israel, whether it is produced locally or transported through their territorial waters.[5]

In accordance with the previous recommendations, a permanent Petroleum Office was established on 17 July 1954, under the supervision of the League's Political Committee. On 15 January 1959 this office became the Department of Petroleum Affairs. One of the major activities of this department has been the sponsorsing of Arab Petroleum Congresses. As the Secretary-General of the Arab League said, the real target of the Arab Petroleum Congress was:

the achievement of closer co-operation between member states in the economic, financial, cultural and other fields . . .The idea underlying this Congress was purely cultural in its origin, and consists in disseminating a general knowledge of petroleum affairs among the Arab peoples.[6]

In association with the Department of Petroleum Affairs, the general objectives of the Arab Petroleum Congresses were twofold:

1. Promotion of Petroleum Knowledge among the Arab Countries

One of the main objectives of holding an Arab Petroleum Congress annually and a Petroleum Exhibition every four years was to promote petroleum knowledge among the Arabs and create a petroleum consciousness. The Department of Petroleum Affairs prepared memoranda, documents and research reports needed for the meetings of Arab officials dealing with oil matters.

In the past, Arab governments had been totally dependent on the major companies for the economic information and analyses used in formulating oil policies in the Arab oil-producing states. As late as 1951, for example, the Iraqi negotiators neither knew anything about the function of the posted prices on which their country's revenues were to be calculated, nor did they have access to economic studies that would help them formulate a negotiating position *vis-à-vis* the oil companies. Later, the publication of the Federal Trade Commission study on the international petroleum cartel provided some factual bases and analytical insight for many Arab nationalists trying to understand the operations of the oil industry in the Middle East.

2. Co-ordination of the Petroleum Policies of the Arab Countries

Keeping in mind the terms of the old oil concessions and the attitudes of the Majors' governments toward the Arab countries, one of the League's principal tasks was to achieve 'the close co-operation of the member states' in economic and financial affairs. Oil was the predominant national resource for many of the League's members. When they achieved political independence in the post-war period, each country reacted differently according to its particular strategy for acquiring a greater voice in its oil affairs. As mentioned earlier, some called for the immediate nationalization of all oil companies, while others preferred to co-operate with the Majors with an eye to more Arab representation in the companies' decision-making process.

The Arab League worked to co-ordinate the diffused efforts of its members in the matter of nationalization. While nationalism was dear to the Arab people, it was commonly acknowledged by most Arab experts that nationalism represented an unrealistic approach to oil matters. The scarcity of skilled administrators and technicians in the Arab world presented a major obstacle to effective nationalization, notwithstanding the political difficulties involved. Recognizing these limitations, the majority of Arab producing states agreed on a more realistic approach through co-operation with the Majors to achieve more favorable terms. This strategy was strongly recommended by the Arab League. The Second Petroleum Congress suggested some general rules for the Arab producers in dealing with the oil companies, among them that the Arabs form a united front and thereby negotiate from a position of strength. The principal resolutions of the Congress were these:

(a) The Congress requested an improvement in the terms granting oil concessions, and expressed its hope that the companies would respond favorably to such equitable demands in order to ensure the continuation of fruitful co-operation between the two parties.
(b) The Congress expressed dissatisfaction with the companies' unilateral action in reducing the prices of both crude oil and its other products without prior consultation with the Arab oil-producing states.
(c) The Congress recommended that the Arab countries double their efforts to fund and absorb various technical, economic and legal studies — all of which would build a corner-stone for better expertise in oil affairs.[7]

One form of co-operation advanced by the Congress was to establish national petroleum companies to be the nuclei of the future Arab petrochemical industries. At its fifth meeting, the Congress recommended that the governments of the Arab states take measures to establish an Arab Petroleum Organization, an Arab Petroleum Research Institute and an Arab Petroleum Company to be operated collectively within the framework of the Arab League. Again, the immediate objectives were closer co-operation in formulating common petroleum policies and training Arab researchers in the various sectors required by the petroleum industry, in the hope of initiating economic development and consolidating the economies of the Arab states. These recommendations were further implemented during the third meeting of the Arab Petroleum Congress in Alexandria from 16 to 21 October 1961.[8]
Although the major objectives of the Arab Petroleum Congresses

are non-political in nature, one of the most important political pre-occupations of the Arabs is the Palestine problem. The creation of Israel resulted in one and a half million Palestinian Arab refugees, and permanent tension in the Middle East. Thus the Arab Petroleum Congresses could not overlook the main preoccupation of the Arab masses and Arab policy in general.

From the beginning, some Arab leaders, supported by the wave of nationalist feeling, tried to emphasize the political goals at the expense of the economic ones and to bring oil under the League's political wing:

Had they succeeded, the League would have become a significant factor in world affairs. The countries which had oil, however, stubbornly and consistently refused to allow it to become a political football of Arab politicians. Although the League had a committee on oil, it never functioned because the oil producing centers − Saudi Arabia and Iraq at the time − jealously kept the Arab League leadership, Egypt, out of the oil business.[9]

It was not until the Fifth Arab Petroleum Congress, held in Cairo from 16 to 23 March 1965, that a major cry for the use of oil as a political weapon was voiced. The tone of the Congress was much more rhetorical and political than any of the previous Congresses. At the opening of the Congress, Ahmed Kamel al-Badri, Chairman of the Congress, declared:

Arab petroleum today is, as it has ever been, the axis and object of all the conspiracies plotted by the alliance of colonialism and Zionism . . . The weapon that Arab petroleum represents can be redirected toward the heart of Zionism and colonialism, should they ever be tempted to commit any new acts of aggression.[10]

The Congress recommended that oil and other potential economic weapons be used effectively in the struggle of the Arab nations to liberate Palestine, and that the relations of the Arab states with all other states be determined in the light of their attitudes to the cause of Palestine. The Congress also endorsed the basic demands of OPEC. By 1967 the use of oil as an indirect weapon against Israel had taken firm root in Arab minds. Thus, within the framework of the Arab League and among Arab nationalists around the Arab world, the twin issues of oil and politics had coalesced about the Palestine question.

Notes

1. Quoted in League of Arab States, *Efforts of the League of Arab States in Petroleum Affairs, 1945-1965* (Petroleum Affairs Department Publication, 22 March 1965), pp. 7 and 8.

2. Joe Stork, *Middle East Oil and the Energy Crisis* (Monthly Review Press, New York, 1975), p. 211.

3. Zuhair Mikdashi, *The Community of Oil Exporting Countries* (George Allen & Unwin, London, 1972).

4. George Tomeh, 'OAPEC: Its Growing Role in Arab and World Affairs', *Journal of Energy and Development*, vol. III, no. 1 (Autumn 1977), p. 28.

5. League of Arab States, 'Efforts in Petroleum Affairs', p. 6.

6. Quoted in ibid., p. 27.

7. League of Arab States, *Fifth Arab Petroleum Conference Resolutions No. 4, 5 and 6* (Dar al-Hana Press, Cairo, 1965), p. 88, 8/A and 8/B.

8. League of Arab States, *Resolutions and List of Papers of First, Second, Third and Fourth Arab Petroleum Congresses*, Resolution no. 1 (Mondiol Press, Cairo, 1964), p. 4.

9. Benjamin Shwadran, *The Middle East, Oil and the Great Powers*, 3rd edn (John Wiley & Sons, New York, 1973), pp. 399-500.

10. Ibid., p. 504.

PART THREE
THE POLITICIZATION OF ARAB OIL

9 THE POLITICIZATION OF ARAB OIL

It is more than four decades since the Arab states first recognized the potential of their oil and its strategic position in world affairs. During the historic meeting with President Roosevelt during the Second World War, the late King Abdulaziz ibn Saud of Saudi Arabia said: 'Whoever controls the still untapped oilfields of the Middle East will have power to make peace or war.'[1] President Nasser, in his book *The Philosophy of the Revolution*, described oil as one of the three chief components of Arab power. 'Petroleum', Nasser asserted, 'is the vital nerve of civilization, and without it civilization cannot possibly exist'.[2] Using oil as a political weapon against Israel and its supporters is an old theme. As Seymour Brown has said:

> [F]ew countries in the Third World have natural resources such as oil which can be used directly as a bargaining weapon in their confrontation with the industrialized world or indirectly through accumulation of financial assets to strengthen bargaining demands.[3]

The Arabs realize that oil exploitation constitutes a substantial foreign element in the midst of a society whose main objective has been, and continues to be, to free itself from the influence of external powers.

The history of oil is interwoven with the modern history of the Arab world. Because of the protracted struggle over oil concessions by foreign powers in the Arab world and the adverse impact of the concessions on Arab economy and society, the term 'concession' has unfavorable connotations. Indeed, its use has been condemned as derogatory to national honor, because of the close association of the concessionaire companies with Western imperialist regimes, which are colonialist in practice:

> The Arabs look at their oil — in spite of the enormous profits they gain from it — with suspicion and distrust, and they link the oil companies that exploit this oil with the history of imperialism in their countries. In doing so, they are not far from reality, as Britain once considered its military base in Aden a must for the protection of the oil which was exploited by British financial investments in the Arabian Gulf.[4]

93

The modern history of oil in the Arab world coincides with the drive of major international oil companies to co-ordinate their oil policies towards the region. In 1928, backed by their home governments, the major oil companies signed the Red Line Agreement to divide the oil areas in the Arab world. According to this agreement, each participant in the Turkish Petroleum Company (TPC), later the Iraq Petroleum Company (IPC), would refrain from taking separate action in an area which included most of the old Ottoman Empire — marked by a red line on a map of the Middle East. This area was to be developed only through joint action by all the participants. In the words of Raymond F. Mikesell:

> This agreement is an outstanding example of a restrictive combination for the control of a large portion of the world oil supply by a group of companies which together dominate the world market for this commodity.[5]

The Arab drive for unity was usually viewed by Western nations as a threat both to their oil supplies and to their domination over the area, and was met by the most efficient colonial method of 'divide and rule'. Thus, to assure Western oil supplies on 'favorable terms', the Arab world was divided into small, artificial states and principalities, some of which have populations of no more than 5,000. The cartel of 1928 was an important feature of the 'divide and rule' doctrine. As Professor Hamed Rabie, a distinguished Arab author wrote:

> This cartel, which was a reflection of the Western, particularly Anglo-Saxon, capitalist interests was influenced by one spirit and was guided by one principle, expressed by Great Britain's representative in the area in a famous telegram in April 1920: 'I recommend that we keep the Iraqi region under the absolute control of Great Britain. We must not let it unite with the other parts of the Arab world. On the contrary, we must work to isolate it by all possible means from the other parts of the Arab nation.'[6]

Keeping all these factors in mind, it should not be surprising that political considerations have shaped contemporary Arab political philosophy on oil. To the Arabs, the 'oil problem is political before it is economic'. After the First World War, the Arabs used the words 'imperialism' or 'colonialism' to describe Western policy towards the Arab world. Such labels were used by the Arabs because of bitter memories and long experience. Arab participation in the First World War, side by side with the Allied forces, was met not only by the division and occupation of Arab lands, but also by the Balfour Declaration which promised the most strategic portion of the Arab world to the international Zionist movement. Thus the major oil companies (all Western) operating in Arab lands were labelled by some Arab nationalists as agents or tools of Western imperialism. 'It is no secret that these companies appeared on the Arab stage at a time when Western policy was very eager to extend its grip on the Arab world.' As rightly described by Peter R. Odell:

> Oil companies . . . became the 'whipping boys' for the 'colonial' powers, which may no longer have direct political control but which certainly exercise significant influence through their control over the purse strings.[7]

Thus when the Arabs finally achieved political independence in the 1950s, they found it impossible to dissociate the oil question from political considerations. Their major objective was to rid themselves of Western imperialism. However, their reactions differed from country to country. Some wanted to follow the Mexican example of 1938 by nationalizing all foreign oil companies, while others preferred to co-operate with the companies provided they had greater voice in their oil affairs.

There were, in fact, three different historical developments that converged to bring about the oil-producing countries' final control over their natural resources. These developments were: first, structural change in the oil industry; second, the national liberation movements and revolts in the 1950s and 1960s; third, the increased dependence of the world on oil. Apart from these common factors, two particular factors had considerable influence on the Arab oil producers and contributed directly to the politicization of oil and its use to effect national objectives. The first pertains to the Arab–Israeli conflict, whose roots are anchored in the creation and subsequent expansion of Israel; the second concerns the dynamic nature of inter-Arab politics.

Although Arab leaders were unable to wield the oil weapon effectively in the early days of its discovery, they were not oblivious to its strategic significance and potential, as they witnessed the world powers contending for control over oil reserves. The Arabs experimented with the use of the oil weapon on several occasions before the Ramadan war of 1973, but it was then that oil was used to its maximum efficiency. In what follows, we will first trace the early attempts at using oil tactically in the confrontation with Israeli expansionism, and then identify the factors that contributed to the successful use of oil in consolidating Arab military victories.

Notes

1. Ernest Jackl, *Background of the Middle East* (Cornell University Press, Ithaca, N.Y., 1952), p. 13.
2. Gamal Abdul Nasser, *The Philosophy of the Revolution* (Ministry of National Guidance, Cairo, 1954).
3. Seymour Brown, *New Forces in World Politics* (Brookings Institution, Washington, D.C., 1974), p. 93.
4. Mohamed T. al-Ghoneimi, *Al-Petrol al-Arabi wa Azmat al-Shark al-Awsat* (Arab Oil and the Middle East Crisis) (in Arabic) (Cairo, 1974), p. 118.
5. Raymond F. Mikesell and Hollis Chenery, *American Oil: America's Stake in the Middle East* (University of North Carolina Press, Chapel Hill, 1949), p. 45.
6. Hamed Rabie, *Arab Oil and Liberation Strategy* (Arab Awakening House, Cairo, 1971), p. 32.
7. Peter R. Odell, *Oil and World Power* (Penguin Books, New York, 1979), p. 47.

10 PREVIOUS ATTEMPTS TO USE OIL AS A POLITICAL WEAPON

The Arab boycott during the Ramadan war in October 1973 was not the first time that the Arabs had used oil as a political weapon. In 1948, 1956, and again in 1967 during the Six-Day war, the Arabs had tried many times to use their ultimate weapon, but their attempts were thwarted by the hard economic and political realities of the time and by the political indecision of their leadership. These previous incidents passed virtually unnoticed insofar as their economic and political impact on the Western oil-consuming countries was concerned, but they were of great consequence in inter-Arab politics. As events later proved, it was the Suez crisis which triggered Arab action and hardened the Arabs' attitude towards the West. The idea of using oil as a political weapon matured with every major issue and external threat that confronted the Arabs, and oil tactics grew more effective with time.

The high-water mark of Western power and control in the oil-producing countries of the Middle East came in 1953, the year of the Anglo-American coup which denationalized the Iranian oil industry and restored the Shah to his throne. The nationalization venture had been a failure, but in a sense, it was 'a harbinger of things to come. For the oil industry it was a victory but in some way it also marked the beginning of the end.' Political changes occurred which, as early as the 1960s, weakened the foundation of Western power and control over the Middle East oil industry.

The Iranian nationalization act was a great challenge to Western control. It took the combined power and intrigue of the United States and Britain to reverse it and stabilize the situation. However, the seeds of change were already germinating. In July 1956 Western powers were shocked by Nasser's nationalization of the Suez Canal Company. The situation in the Middle East was not so stable after all. The West's reaction was one of panic and confusion. However, the issue of the nationalization of the Canal was not the primary cause of Western concern:

It was obvious that behind the more limited and immediate Suez Canal issue, as well as behind the general East–West crisis in the Middle East, there lurked the question of the fabulous oil reserves of

97

the region. The oil is frequently given as both a cause and a reason for the crisis, and the oil is raised like a talisman to silence any discussion and to answer any criticism against given policies and practices.[1]

With nationalization, the West European nations, whose economies depended to a large extent on Middle East oil and oil products, sensed an immediate threat to the security of their sea-borne commerce. It was estimated that 'the United Kingdom had reserves of oil which would last for six weeks, and that the other countries of Western Europe owned comparatively smaller stocks'. Nationalization of the Canal was regarded as a challenge to the entire Western politico-economic conceptual framework for the Middle East.

After Nasser's drastic step, the negotiations on the future of the Canal were brought to an abrupt end with the tripartite Anglo–French–Israeli attack on Egypt. On 31 October 1956 the Egyptians blocked the Suez Canal and on 3 November the Syrian Army blew up several IPC installations, which cut off the flow of oil from Iraq to the Mediterranean. The Kirkuk-Tripoli pipeline was 'the biggest single piece of Middle East production that went out'. These unprecedented actions created a crisis without parallel in the history of the Middle Eastern oil industry, and the effects of these interruptions were worldwide.

Without going into detail, it is sufficient to say that the United States played a major role in lessening the most severe effects of the oil shortage on its West European allies. The Americans organized a worldwide reallocation of crude oil, oil products, and oil-carrying tanker fleets to rescue Western Europe. The 'Emergency Oil Lift to Europe . . . was perhaps second only to the Marshall Plan in the grandiose scale of its design and the magnitude of its operation'.[2]

The events which followed the 1956 Suez crisis failed to produce any serious effects on Western industrialized nations or to advance significantly a united Arab strategy for the use of oil as a political weapon. But its future impact could not be easily ignored. In 1967, for a second time, the Arabs used oil against the West. The Suez Canal was again closed, this time for a long period. The Arab producers' decision to stop production was carried out, even though they themselves doubted its practicality. Nevertheless, the Arabs knew that European dependence on Arab oil had nearly tripled since 1956, and some Arabs were convinced that success was imminent. They argued that Europe was absolutely dependent on Arab oil and that European industrial production would be paralyzed if its oil flow were shut off. Similar industrial misfortune was predicted for the United States. This view was advanced by the 'progressive' countries, mainly Egypt, Syria, Iraq and Algeria. In June 1967 the Algerian Foreign Minister said: 'The West needs Arab oil much more urgently than the Arabs need to sell it to the West.'[3]

However, the Arab boycott was again short-lived. The oil producers quickly realized that they were hurting themselves more than their Western customers. The end of the 1967 war left the Arab producers much worse off than before. Iran and Venezuela, the non-Arab members of OPEC, had taken advantage of the boycott to increase their exports into Western Europe, and even Libya was able to export more to West Germany. The United States, which was expected to suffer most from the embargo, had hardly been affected. Thus, despite the growing politicization of Arab oil, the effects of the embargo in 1967 were scarcely more severe than those in 1956. As Christopher T. Rand stated:

In spite of the differences between these two points of time, however, one can argue that the Arab oil cutoff was little more severe in 1967 than in 1956: that although the Arabs had become far more aware of the power of their weapon during that decade than they had been before, they had not in general become more convinced of its utility.[4]

Sheikh Ahmed Zaki Yamani admitted that the use of the oil weapon had been a fiasco: 'If we do not use it properly,' he said, 'we are behaving like someone who fires a bullet into the air, missing the enemy and allowing it to rebound on himself.' The second major embargo imposed by Arab oil-producing countries did not produce the desired effects. There are various reasons for this failure:

(1) The US, the prime target of the Arab oil weapon, was immune to the Arab embargo because it was at that time totally self-sufficient in terms of Western hemisphere oil supplies.
(2) The international oil companies did an outstanding job of making up the shortfall in Arab oil supplies to the embargoed countries from other sources, despite the closure of the Suez Canal. In this case, the focus of the world oil supply crisis shifted somewhat from the availability of crude oil at the source to a tight situation as regards transportation.

(3) No quota ceilings were imposed on production, with the result that no actual physical shortage of oil was created.

(4) There was no uniform interpretation of the coverage of the embargo. Some of the North African oil companies did not in fact withhold supplies from West Germany.

At the Arab Summit Conference in Khartoum in August 1967, the Arab states gathered to resolve their differences and to formulate a united Arab policy towards Israel and the West. The Arabs agreed on a formula for bringing the fighting between the conflicting forces in Yemen to an end and also on a new philosophy for the use of oil. The oil-producing countries agreed that, instead of cutting back their oil supplies to the West as advocates of the 'oil weapon' had been urging, the new policy would be to make deliberate use of oil revenues to provide the financial basis for the achievement of political ends. In brief, they would give financial support to the countries which had suffered from Israeli aggression, 'since oil is a positive Arab resource that can be used in the service of Arab goals'. The expanded scope of this new policy initiative emphasized oil as a creative, dynamic force in the Middle East, and emerged logically from the growth in total exports and the growth in revenue per barrel, which had risen from about 30 cents in 1946 to about 80 cents in 1966.

Nevertheless, it was clear that the Arab producing countries were not yet entirely united, and that the moderates — Saudi Arabia, Kuwait and pre-revolutionary Libya, which were solely dependent on oil for their revenues — believed they were being coerced into using oil as a political weapon by radicals who were less dependent on oil revenues, among them Egypt, Syria and Algeria. In 1967 the moderates took the initiative in steps leading to the establishment of the Organization of Arab Petroleum Exporting Countries (OAPEC). Saudi Arabia, Kuwait and Libya argued that the original membership should be deliberately confined to those states in which oil 'constitutes the basic source of . . . national income'. When OAPEC was formally established in 1968, the only independent country other than the founders that could meet this requirement was Iraq, but Iraq expressed reluctance at identifying with the other three regimes in a 'club of conservatives'.

The first meeting of OAPEC was held in Kuwait from 9 to 11 September 1968. Membership was open only to those approved by three-quarters of the votes of the founding members, with each of the founders having a veto. OAPEC declared that it was to be a non-political organization. Its objectives were regarded as fully compatible with those of OPEC, although the purpose of OAPEC was to have a separate Arab organization to co-ordinate the achievement of purely Arab, as distinct from OPEC's multi-national objectives. However, unlike OPEC which was concerned mainly with 'the co-ordination of members' policies in one commodity, petroleum, and solely in the export market', OAPEC had, as its ultimate goal, the integration of its members' national economies into a single regional economy — 'the EEC of the Arab oil producers', as Sheikh Yamani has described it.

After the revolution in Libya in 1969, OAPEC offered membership to other Arab oil producers. In May 1970 Algeria, Abu Dhabi, Bahrain, Dubai and Qatar were formally admitted to OAPEC. In June 1971, after a serious dispute over Iraqi membership, Iraq was finally admitted. Inevitably, as it expanded, OAPEC took on political dimensions, and it was OAPEC which initiated and organized the oil war of October 1973.

Notes

1. Benjamin Shwadran, 'Oil in the Middle East Crisis (V)', *Middle Eastern Affairs*, vol. 8 (April 1957), p. 126.
2. Shoshana Klebanoff, 'Oil for Europe: American Foreign Policy and Middle East Oil', unpublished PhD dissertation, Claremont Graduate School, 1973, p. 251.
3. John H. Lichtblau, 'The Politics of Petroleum', *The Reporter* (13 July 1967), p. 26.
4. Christopher T. Rand, *Making Democracy Safe for Oil: Oilmen and the Islamic East* (Little, Brown & Co., Boston, Mass., 1975), p. 87.

11 THE ARAB OIL REVOLUTION, 1970-1973: REVENUES, PRICES AND AGREEMENTS

Within two weeks of the outbreak of the Arab-Israeli war in October 1973, OAPEC members met in Kuwait. They decided to cut back oil production and exports by a minimum of 5 percent a month in relation to September 1973 levels, and to ban oil shipments to the United States until: (1) that country modified its pro-Israeli stand; and (2) the Israelis withdrew from all Arab territories occupied during the June 1967 war. Only Iraq dissociated itself from the production cutbacks, although its solidarity with OAPEC's objectives was clear from its prior act of nationalizing American interests in IPC.

The decisions endorsed by OAPEC in Kuwait were drastically different from those taken in Khartoum after the June 1967 war. The changes in outlook from Khartoum to Kuwait were due to major structural developments in the international oil industry. Between 1970 and 1973 a series of negotiations had started between the oil companies and OPEC members. John Root, Manager of Exxon, called them 'the most wide-reaching oil talks since World War II'.

Until 1970 OPEC's achievements in terms of price increases had been minimal. At the Second International Symposium on Energy, held in Rome from 11 to 13 March 1968, OPEC's Secretary-General Francisco Parra evaluated OPEC's efforts in the following manner:

> Some price weakness still persists in spite of the fact that OPEC effectively stabilized posted crude prices and inhibited the downward slide in market prices . . . This price weakness tends to have detrimental effects on OPEC member countries, both with respect to the establishment of more equitable levels of tax reference prices, as well as the returns which may be earned by national oil companies entering the market.[1]

OPEC had succeeded only in preventing any reductions in posted prices since 1960, and in reaching an agreed time-table for the full expensing of royalties and the elimination of marketing allowances.

The first sign of upheaval was Libya's changing its posted prices in September 1970, which in turn led to the development of the Tehran and Tripoli agreements. Much of Libya's political significance today is

102

rooted in its emergence in the 1960s as a major crude oil producer. Libya was seen by Western oil strategists as an alternative to the 'unstable' political environment that characterized the rest of the Arab world in the 1950s and 1960s. West European reliance on Libyan and North African oil increased with the rising tension in the Middle East during the mid-1950s. Production climbed sharply in 1957 and 1958 after the nationalization of the Suez Canal, and after the American and British military intervention in Lebanon and Jordan respectively following the 1958 revolution in Iraq. The June war of 1967 resulted in the closure of the Suez Canal and also in a world oil shortage. These events put Libya in a favorable position, and oil production in Libya increased greatly to meet the European demand. By 1968 Libyan oil production had increased by a staggering 49 percent, while the increase in Middle Eastern production was only 12 percent. At 2.8 million barrels at the end of the year, Libya had overtaken Kuwait as the third largest Middle Eastern producer, behind Iran and Saudi Arabia. Another 22.9 percent jump in production in 1969 put Libya's daily production at an average of 3.1 million barrels per day (b/d). By that time Libya was supplying Germany with 45 percent of its requirements, Italy with 28 percent, the United Kingdom with 22 percent and France with 17 percent.

Pushed by the dynamics of competition between the Major and Independent oil companies and free from any political interference from the Libyan monarchy, prices for Libyan crude remained solid as the market for Libyan oil expanded. Nevertheless, encouraged by Libya's advantageous position, King Idriss demanded a 10 cent hike in the posted price for his oil. When the oil companies hesitated, the King gave them an ultimatum either to raise their posted prices to levels assessed by the government or to be ready to face unilateral action. The deadline to comply with the ultimatum was 1 September 1969. On that day, a revolution brought young army officers to power in Libya.

Tripoli: Phase I

Up to this point none of the producing countries, either singly or collectively through OPEC, had made any attempt to increase their revenues by increasing oil prices. The new Libyan regime, meanwhile, announced that Libya would honor its oil concessions and that, although 'no spectacular changes' would be made, its policy would include 'more effective control' over oil production. The Libyan Oil

Minister Ezzedin al-Mabruk expressed his government's oil policy in the following manner:

> We do not wish to dig up the past, nor to bring it back to mind. What we wish to emphasize, with absolute clarity and frankness, is that the new revolutionary regime in Libya will not be content with the previous passive methods of solving problems . . . The just demands we seek here are not intended to bring about any basic changes in the existing structure of the world oil industry, nor specifically in the price system. This does not mean to say that we approve of the existing system or believe that it is an equitable one. Libya will continue to support the collective efforts undertaken by OPEC to alter conditions in this respect . . . What we ask of you, gentlemen, is to recognize the changed circumstances in our country and, accordingly, let your actions be guided by flexibility and reasonableness.[2]

However, this did not give any comfort to the oil companies and the consuming countries. What they feared would happen did indeed happen a few months later. In January 1970 Libya entered into negotiations with Esso and Occidental, representing the Majors and the Independents respectively. It was, in fact, a continuation of the previous regime's attempt to raise the tax reference price of Libyan crude. However, although the companies seemed unwilling to grant a 10 cent hike, for two months they pondered legal action over the government's request, in essence rejecting it through delay. The Libyan government became increasingly frustrated with the companies' continued postponement of their reply. In April, former Prime Minister Maghrabi was put in charge of negotiations and announced that the 10 cent rise was not enough. Meanwhile, the deadlock persisted. Taking a new tack, the government, after presenting its demands to all companies, focused its attention on Occidental, which concentrated most of its oil production in Libya. In May, the Libyan government ordered Occidental to cut its oil production by 300,000 b/d. A month later, the Oasis Consortium (three American Independents) was trimmed by 125,000 b/d. By mid-August 1970, the government had hit Esso with a 110,000 b/d cut, Mobil with a reduction of 55,000 b/d, and ordered another 60,000 b/d cut from Occidental. Libyan production was down by almost 800,000 b/d. In addition to these cutbacks, the 'accidental' damage to the Trans-Arabian Pipeline (Tapline) in May resulted in the stoppage of around 500,000 b/d of Saudi oil to Western Europe. Syria, in alliance with Libya, refused to allow the company to make the needed repairs. These moves were unprecedented in Middle Eastern history.

As the first anniversary of the revolution approached, the Libyan officials, recognizing the strength of their position, concentrated their negotiating pressure on Occidental and Esso. On 30 August 1970 Occidental buckled, reportedly under the threat of take-over. Posted prices were raised as of 1 September 1970 by 30 cents per barrel, with a further rise to $2.55 per barrel as of 1 January 1971. Taxes were raised to 58 percent as opposed to the previous 50–50 sharing formula. The government immediately restored Occidental's full production rate. Libya then told the other companies to 'take it or leave it'. One by one, the companies capitulated. Some of the tax rates varied, but the basic rate for government profit in Libya was now firmly established at 55 percent.

This was only a prelude to the world-wide wave of bargaining which began when other OPEC members started to follow Libya's lead. As Marwan Iskander put it: 'There is little room for doubt that Libya's success started the snowball rolling and led to Resolution XXI.120 and consequently the Tehran and Tripoli agreements.'[3]

The major oil companies were aware that the Libyan settlement would prompt a new wave of demands by the other oil-producing countries. Even before the new Libyan accords were signed, the Iranian Prime Minister demanded an increase in Iran's income tax to 55 percent and an increase in posted crude prices. Then Saudi Arabia and the Arabian Gulf countries successfully insisted that their share of profit be raised. Within a month, the Majors had agreed to a 55 percent income tax rate in all the Middle East producing states. The question of posted prices was postponed until after the semi-annual OPEC meeting to be held in December 1970.

Between 9 and 12 December 1970 the Twenty-First OPEC Conference met in Caracas, Venezuela, and adopted Resolution 120, which recommended the following principles:

(1) the establishment of a 55 percent minimum rate for taxation of company net income;

(2) a uniform posted price in all member countries equal to that of the most favored countries;

(3) a uniform increase in posted prices in order to reflect favorable trends in world market prices;

(4) the adoption of a new system for the calculation of quality and

location differentials affecting market prices;
(5) the elimination of all discounts offered on crude export prices as of 1 January 1971.[4]

To realize these ends, the conference decided on a negotiating strategy designed around three regional groupings: the Arabian Gulf countries, the Mediterranean exporters, and Venezuela and Indonesia. The conference also declared that its members were ready to legislate for the new terms (to the major oil companies) if negotiations did not proceed to a satisfactory conclusion. The resolution also established a time-table of approximately three to four weeks within which the negotiations were expected to be completed. Furthermore, OPEC threatened that there would be 'concerted and simultaneous action by all member countries' to enforce their objectives if they were not attained according to schedule. As Ruth S. Knowles wrote:

> The OPEC resolutions passed in Caracas left no doubt that the revitalized organization now knew how much muscle it had and was determined to use it. Their objectives were to view 55 percent as a 'minimum' income tax rate, to eliminate all tax advantages the companies had and to establish a 'uniform' general increase in the oil tax prices in all member countries. The revolt of the petroleum-exporting countries was now in full swing.[5]

The first round of negotiations with the companies was set for 12 January 1971, in Tehran. The Gulf negotiating team was to consist of Iran, Saudi Arabia and Iraq. However, on 3 January, even before the Tehran negotiations started, local Libyan representatives of the oil companies were called upon by Deputy Prime Minister Abdul Salaam Jalloud. Libya presented a new set of demands:

(1) a 5 percent hike in the tax rate – as in the Gulf countries – with retroactive claims to be settled either by a cash payment (with a 10 percent discount), or by five-year installments plus interest, or a tax rate above 55 percent;
(2) a post-1967 freight differential and a post-May 1970 freight differential, to be 39 cents a barrel and 30 cents a barrel respectively;
(3) monthly rather than quarterly tax payments;
(4) increased investment in oil and non-oil areas, amounting to at least 25 cents per barrel of exported oil.

If they did not comply with these demands, the companies would be forced to shut down oil production and risk nationalization. This unilateral move by Libya was probably designed to put pressure on its partners in OPEC before the Tehran negotiations commenced.

The US and British governments decided that they could not ignore these developments, and should take an active role in assisting the companies at the Tehran Conference. Representatives of the United States, Great Britain, France and the Netherlands met in Washington to coordinate the companies' strategy with the backing of the Western governments. According to Professor Jean-Marie Chevalier:

> The companies were given government authorization to act collectively, which was, in theory, contrary to anti-trust legislation. This meant that the cartel was made official and given the blessing of the American, British, French and Dutch governments.[6]

The outcome of these meetings was that the industry issued a joint 'Message to OPEC', signed by Standard Oil of New Jersey, Standard of California, BP, Gulf, Mobil and Texaco. Cleared by the Justice Department on 13 January 1971, it was sent to OPEC governments on 16 January. The message proposed the following points as a basis for a settlement:

> (1) revision of posted prices in all OPEC countries, with provision for moderate annual adjustments to take into account 'world-wide inflation or similar criterion';
> (2) new temporary transport adjustments covering the shipment of Libyan crude and other crude oils carried over short distances, which were liable to vary in price depending on freight rate fluctuations;
> (3) no retroactive payments or new increases in the tax percentage level beyond current levels and no new obligatory investment.

The message also expressed 'great concern' over OPEC's 'continuing series of claims' and proposed 'an all-embracing negotiation' between the companies and OPEC. The companies asked for OPEC's reaction as soon as possible, and offered to meet OPEC representatives 'whenever and wherever' the producing governments wished. Each side had made its position clear; the way for negotiation was open.

The Tehran Conference convened as scheduled on 12 January in a very tense atmosphere. The first meeting broke up without any fruitful results. The OPEC oil ministers were frustrated by the companies' position but resolved to stand firm. The Saudi Oil Minister Yamani warned after the 12 January breakdown: 'I am afraid that they are going to have to pay a heavy price for this for it will hurt them as well as the innocent consumer.'[7] The companies' representatives, on the other hand, were alarmed at the deteriorating situation. It became clear that they were stalling for time while a strategy was worked out in New York for dealing with the Tehran negotiations and the Libyan demands. John McCloy, the companies' legal adviser, suggested that, in addition to the companies' general strategy, 'it would be wise if the government could enter into this thing and get the heads of the countries involved to moderate their demands'.[8] To accomplish this task, the State Department sent Under-Secretary John Irwin to Iran. When Irwin arrived to meet the Shah in Tehran, the companies' collective message, with diplomatic support, was received by OPEC and took all OPEC countries by surprise. Denouncing what he called a 'poisoned letter', the Libyan Deputy Prime Minister proclaimed that 'Libya will defeat the consuming countries and also the oil companies.' On 19 January he put pressure on Hunt and Occidental to dissociate themselves from the industry-wide approach 'or face government action'.

In Tehran, meanwhile, the Shah and his Oil Minister, Amuzegar, objected to OPEC-wide negotiations. When Under-Secretary Irwin and the American ambassador to Iran, Douglas MacArthur II, met the Shah on 18 January, they were told of the Shah's position. MacArthur, a well-known supporter of the Iranian position, convinced Irwin to press the companies to agree to 'split the Gulf from the Med'. George T. Piercy, Exxon's Senior Vice-President and one of the companies' two chief negotiators, was surprised by MacArthur's recommendation and called New York to say that it would violate the companies' message. After hurried consultations, the companies compromised and declared:

> We should prefer, and should have thought it would be beneficial, in the interests of time, that the negotiations should be with a group representing all the OPEC members. Nevertheless, we should not exclude that separate (but necessarily connected) discussion could be held with groups comprising fewer than all OPEC members.[9]

Thus the companies recognized OPEC's demands for two separate negotiations, but they nevertheless tried to maintain a bridge between the separate Gulf and Mediterranean discussions. On 25 January the companies split their negotiating team, with BP's Chairman Lord Strathalmond heading the Tehran group and Exxon's Piercy leading the

Mediterranean team.

On 28 January Strathalmond began the Tehran 'Gulf only' negotiations. The Libyans refused to negotiate in Tripoli until the Tehran negotiations were concluded. The Gulf countries set a five-day deadline by which negotiations were to be finalized. The companies proposed an increase in the posted price of 15 cents per barrel and allowances for inflation. The Gulf countries demanded an extra 54 cents a barrel and a much higher inflation allowance. The producing countries threatened to legislate for their terms unilaterally if the companies did not 'voluntarily' accept OPEC's demands; and if the companies did not accept legislation, OPEC threatened to 'take appropriate steps including total embargo'. Only Indonesia abstained from the total embargo vote.

The Tehran Agreement

The Tehran agreement was signed on 14 February 1971. The financial terms contained the following points:

(1) total tax rates on income to be stabilized at 55 percent.
(2) a uniform increase in posted prices for the Gulf states by 35 to 40 cents per barrel, depending on the quality of the crude;
(3) elimination of all previous discounts worth 3 or 4 cents per barrel;
(4) an immediate increase in government revenues by a total of 27 cents per barrel from the current average of $1 per barrel, and rising still further to 54 cents per barrel by 1975;
(5) stability of taxation and basic posted prices for a five-year period.

After concluding this agreement, the total increase in revenue to the Gulf states was calculated to be $1.2 billion in 1971 and $3 billion in 1975.

The Tripoli Agreement

Four days after the Tehran agreement, talks began in Tripoli. The Libyan team was headed by Deputy Prime Minister Abdul Salaam Jalloud, who commented that the Tehran agreement did not meet 'our minimum demands'. In a display of solidarity, the representatives of

Libya, Algeria, Iraq and Saudi Arabia met in Tripoli on 23 February to define a common basis for negotiation. Libya also obtained the support of Syria and Nigeria, which announced their intention to join OPEC.

On 2 April, six weeks after the Tehran agreement, Libya signed the five-year Tripoli agreement with 15 oil companies. In addition to the standard features of the Tehran agreement, the Tripoli agreement included: 55 percent tax rate, five-year guarantee, 2.5 percent inflation allowance, and an increase in the posted price by 90 cents per barrel. The Libyan demand for a specific investment formula was left to be worked out on a company-by-company basis. The cumulative package gave Libya $3.9 billion in additional revenues over the next five years. All countries concerned with Mediterranean crude oil production signed agreements of a similar nature within the next few months.

As a major consequence of the Tehran-Tripoli settlements, for the first time in the public mind, Middle East oil and Middle East politics were linked to the 'energy crisis'. To be specific, the Tehran agreement stated that, if the international oil companies failed to comply with legal and legislative measures adopted by OPEC countries within a week from the date of their introduction, all OPEC members (except Indonesia) would 'take appropriate measures, including a total embargo on the shipments of crude oil and petroleum products'. Thus, for the first time since the June war of 1967, the question of an oil embargo was advocated by the producing countries. However, this time, not only the Arab countries, but also the non-Arab OPEC members agreed to employ the tactic of embargo.

Some Arab writers have linked clauses in the agreements to the Arab-Israeli conflict. Ahmed Baha al-Din, a prominent Egyptian journalist, in an editorial which appeared in *Al-Ahram*, claimed that the companies gave in to OPEC's pressure for higher prices 'as a result of the [1967] war' and fear of 'the atmosphere of Arab hostility toward them as a result of America's support of Israel'. He added that 'the oil companies were anxious to alleviate the probable effects of this wave of antagonism by meeting the demands of the petroleum producing countries'.[10]

Statements by Libyan officials during the negotiations tend to corroborate Baha al-Din's position. Deputy Prime Minister Abdul Salaam Jalloud, for example, in a meeting with oil company representatives, frankly stated that Libya's demands were intended to 'hurt' the companies and force them to put pressure on the United States to change its pro-Israeli policy. To the non-American companies, he emphasized that they must be made to feel the 'effects of Zionist aggression' in order to convince them to press Western governments to change their Middle East policies along lines favorable to the Arab cause. 'When companies suffer,' Jalloud stated, 'they may be impelled to do something to alter United States policy.'

The two agreements in Tehran and Tripoli were expected to hold until 1976. As it turned out, they lasted less than three years. All plans made under the Tehran and Tripoli agreements were overturned by the October Oil Revolution of 1973, during the fourth round of the Arab-Israeli war. The Libyans, in any event, could hardly wait for the ink to dry on the Tripoli agreement to present the companies with a new list of demands. By mid-April 1971 a new agreement was signed in which the Libyan posted price of crude would rise another 80 cents. In response, Lord Strathalmond, the chairman of BP, declared that the industry had become 'a tax-collecting agency'.

At the same time, OPEC was gearing up for a show-down. In January 1972 the Gulf producing countries summoned the oil companies to Geneva to settle two issues: (1) a dispute over revising the Tehran agreement to compensate for devaluation of the dollar and readjustments in other world currencies; and (2) government participation in company ownership. The monetary dispute was quickly settled by an agreement which called for an immediate increase of 8.4 percent in posted prices and adjustments to take into account important changes in the exchange value of the dollar. The increase in revenue to the Gulf states signatories to the agreement was estimated to be $700 million a year. An amendment to the Geneva agreement was signed on 2 June 1973 in Geneva and stipulated an immediate increase of 11.9 percent in the posted price to compensate for the 10 percent devaluation of the US dollar in February; the revised agreement was to be valid until 1975. With the financial issues out of the way, the Gulf states and the Majors turned their attention to the looming issue of government participation in the oil companies themselves.

Notes

1. Marwan Iskander, *The Arab Oil Question*, 2nd edn (Middle East Economic Consultancy, Beirut, 1974), p. 11.

2. *Middle East Economic Survey (MEES)*, 30 Jan. 1974.

3. Marwan Iskander, *Al-Da'm al-Nafti al-Arabi: Min Mu'tamar al-Khartoum 1970 ila Mu'tamer al-Kuwait 1973* (Arab Oil Support from the Conference of Khartoum 1970 to the Conference of Kuwait 1973) (in Arabic) (Middle East Economic Consultancy, Beirut, 1973), p. 78.

4. Between 1960 and 1970 OPEC countries granted official discounts on posted prices; however, it was agreed that these discounts would be phased out by 1975. In this respect, the 21st OPEC Conference only speeded up the phasing out of these discounts.

5. Ruth S. Knowles, *America's Energy Famine: its Cause and Cure* (University of Oklahoma Press, Norman, Oklahoma, 1980), p. 87.

6. Jean-Marie Chevalier, *The New Oil Stakes* (Penguin Books, Harmondsworth, 1975), p. 50.

7. *MEES*, 15 Jan. 1971.

8. Anthony Sampson, *The Seven Sisters* (Viking Press, New York, 1975), p. 218.

9. *New York Times*, 28 July 1974.

10. *Al-Ahram*, 1 Oct. 1972.

12 THE ARAB OIL REVOLUTION, 1970–1973: PARTICIPATION VERSUS NATIONALIZATION

In early 1972 the Gulf members of OPEC and representatives of the international oil companies met in Geneva. The contentious financial issues were resolved quickly, and the OPEC members turned the discussions towards fundamental questions of government participation in the ownership of the oil companies. In an interview with the editor of the *Petroleum Times* in May 1971 Nadhim Pachachi, OPEC's Secretary-General explained the organization's immediate objective:

> We, the producing states, believe that, at some future date, we should become more involved in the downstream operations of the oil industry. The achievement of this objective is currently the major part of the work of OPEC. So far we have made no progress in achieving this objective, but we have been studying the subject fairly strenuously and there is a ministerial committee composed of the Ministers of Iran, Iraq, Kuwait, Libya and Saudi Arabia, who will meet before the next annual conference, which is to be held in Vienna on 12 July, to decide on what form our participation in downstream operation might take. This objective is the most important aim of OPEC, now the posted price issue has been settled.[1]

During this phase of negotiations with the oil companies, Saudi Arabia was to take charge; and Sheikh Ahmed Zaki Yamani, its Oil Minister, was to be the dominant figure. OPEC appointed Yamani, who had been advocating participation since 1968, to negotiate with the oil companies on behalf of all Gulf producing countries. Yamani announced that he would seek 20 percent participation as the 'minimum of minimums' and that compensation would be determined solely by the book value of company investments and would not include any allowance for proven oil reserves for the simple reason that 'the oil is ours'.

Sheikh Yamani explained what the Gulf states sought from participation in a lecture delivered at the American University of Beirut in June 1969, during which he stressed that participation would also be in the interests of the oil companies:

113

It will save them from nationalization, and provide them with an enduring link with the producing countries . . . For our part, we do not want the majors to lose their power and be forced to abandon their roles as a buffer element between the producers and the consumers. We want the present set-up to continue as long as possible and at all costs to avoid any disastrous clash of interests which would shake the foundations of the whole oil business. That is why we are calling for participation.[2]

Yamani began his discussions with Aramco. The company tried to put off participation by suggesting an alternative proposition — that the Saudis could have a 50 percent share in future discoveries. But Yamani flatly rejected the proposal. Eventually, in October 1972, Yamani and Aramco reached a 'general agreement' which provided for 25 percent government participation in established Gulf producing countries, and for gradual escalation to 51 percent ownership in 1983. The final agreement was signed at the end of the year in Riyadh.[3] Table 12.1 shows both the percentages of participation (a), and the percentages of corresponding 'phase-in' (b).

Participation was a relatively new concept for the Arab governments to advance. Historically, nationalization has been seen as the only way to break the hold of a foreign monopoly over a developing country, because its benefits are twofold — eliminating foreign domination and requiring a reorientation of the local economy. Although nationalization had occurred within the oil industry both before and after the formation of OPEC, the producing countries had, generally speaking, avoided nationalization, partly because of the failure of Mossadeq's nationalization of AIOC in Iran in 1953. For some Arab nationalists, the whole concept of oil concessions, granted to the oil companies in earlier times under disadvantageous conditions, was increasingly unacceptable. They argued that the only way to rid the Arab states of these unfair concession agreements was by nationalization.

The Kuwait daily newspaper, *Al-Ra'i al'Am*, reviewed the case for nationalization at the beginning of 1970. After stating that there was some consensus of opinion as to the beneficial effects of nationalization in developing countries, the article commented:

This unanimity arises from the belief that the measure asserts the sovereignty of a nation over its territory and constitutes the best means of securing social justice, in that the resources thus acquired remain in the hands of the producer countries.

Table 12.1: 'Participation' and Corresponding 'Phase-in' (percentages in terms of total production)

Year	Initial (a)	Initial (b)	First increase (a)	First increase (b)	Second increase (a)	Second increase (b)	Third increase (a)	Third increase (b)	Fourth increase (a)	Fourth increase (b)	Fifth increase (a)	Fifth increase (b)	Total (phase-in)
1973	25	3.75											3.75
1974	25	7.50											7.50
1975	25	12.50											12.50
1976	25	17.50											17.50
1977	25	16.25											16.25
1978	25	15.00	5	4.50									19.50
1979	25	12.50	5	4.00	5	4.50							21.00
1980	25	10.00	5	3.75	5	4.00	5	4.50					22.25
1981	25	7.50	5	3.50	5	3.75	5	4.00	5	4.50			23.25
1982	25	2.50	5	3.25	5	3.50	5	3.75	5	4.00	6	5.40	22.40
1983			5	3.00	5	3.25	5	3.50	5	3.75	6	4.80	18.30
1984			5	2.50	5	3.00	5	3.25	5	3.50	6	4.50	16.75
1985			5	2.00	5	2.50	5	3.00	5	3.25	6	4.20	14.95
1986			5	1.50	5	2.00	5	2.50	5	3.00	6	3.90	12.90
1987			5	0.50	5	1.50	5	2.00	5	2.50	6	3.60	10.10
1988					5	0.50	5	1.50	5	2.00	6	3.00	7.00
1989							5	0.50	5	1.50	6	3.00	4.40
1990									5	0.50	6	2.40	2.30
1991											6	1.80	0.60
											6	0.60	

(a) = Percentage of participation.
(b) = Percentage of corresponding 'phase-in'.
Source: Mana Saeed al-Otaiba, *OPEC and the Petroleum Industry* (John Wiley & Sons, New York, 1975), p. 172.

The threat of nationalization could be used as a bargaining tactic or be advanced as part of a genuine program for redressing fundamental economic grievances. The article continued:

Some have used it [nationalization] as a slogan, whilst others have called for it out of conviction and with deep sincerity. Despite the fact that there has not been any positive outcome, it is a burning issue and remains a powerful means of pressure which some manipulators − behind the scenes − do not lack skill in applying.

However, most OPEC members, ever mindful of Mossadeq's disastrous episode, favored the more cautious method of participation. This conservative tendency provoked the emergence of the 'anti-OPEC' group of oil technocrats and economists, whose political program for the Arab oil producers was based on the desirability of nationalizing the foreign companies.

While most of the Gulf states adopted 'participation', Iraq and the Mediterranean states of Libya and Algeria banded together to adopt a common policy, a sequence ending in complete nationalization. The communiqué issued in Algiers on 23 May 1970 by the Oil Ministers of Algeria, Libya and Iraq emphasized:

That the national use of oil resources represents the best weapon for the attainment of economic development and political independence. To this end they declared their determination to strive for the integration of oil operations within the national economy in order to ensure the success of the efforts being undertaken by the three revolutionary powers to fulfill the legitimate aspiration of their people for a better life. The three ministers consider that the best guarantee of success lies in the existence of an authentic national industry in all the principal phases of oil operation, representing the first stage of direct exploitation and full government control over national resources.[4]

There were some examples of successful nationalization which could not be ignored. In Iraq in 1961, the government of Colonel Qasim had promulgated a law which expropriated the IPC concession area, except for those fields under current production (approximately 99.5 percent of the whole concession area). The land nationalized included the North Rumaila fields which belonged to IPC. Over the next several years, there was a sharp political struggle between the government and

IPC over this course of action, and it was not until 1967 that IPC and the Iraqi government came to a partial agreement. In 1972 IPC and the Iraqi government clashed again when the company cut back its oil exports to the Mediterranean; in March and April, production in the Kirkuk oilfields was down 50 percent. Talks began but no progress was made, with the result that the government issued a two-week ultimatum in mid-May, insisting that IPC restore Kirkuk production to normal levels. When the company refused to make a satisfactory response by the end of May, the company's assets were nationalised on 1 June 1972.

Algeria adopted a unique style of nationalization. On 27 July 1965 Sonatrach, the state oil company, concluded a 50/50 partnership agreement, referred to as an 'Association Coopérative' (Ascoop), with Sopefall, a French oil company controlled by ERAP. The agreement ensured for Sonatrach the rights for exploration, production and the sale of oil and gas. Soon after this, the Algerian government decided to place the British and American oil companies in Algeria under government control as a 'punishment' for the 'pro-Israeli policies' of the United States and Britain during the Six-Day war in 1967. In June 1970 Algeria nationalized Shell, Phillips and other small operators. The Algerian Oil Minister outlined the strategy behind his government's oil moves in a speech to a meeting of Arab economists:

> In short, the path . . . is to take in hand the national oil industry and, as much through the many production and service activities related to it as by the cash flows it creates, act so as to integrate it fully in the economic life of the country, and by means of a systematic commercialization of the products derived from it make it a permanent source of income. In other words, it is a question of 'sowing' oil and gas in order to reap factories, modernize our agriculture, diversify our production and create an organized national economy oriented towards progress.[5]

Instead of expropriating the physical assets such as plants, port facilities and pipelines, the Algerians introduced a compulsory take-over of equity shares. Algerian nationalization resulted, therefore, in Sonatrach's becoming the largest shareholder in the Algiers Refinery, and being actively involved in the management of company operations side-by-side with foreign shareholders. Finally, in February 1971, President Boumedienne announced that Algeria was taking a 51 percent controlling holding in French oil companies. When the French partner, infuriated by this act, suggested total nationalization, the Algerian government obliged.

Libya's nationalization was based on a different set of rules. While Iraq and Algeria emphasized economic grounds, Libya's nationalization acts were more politically motivated. The Libyan president, Colonel Gaddafi, expressed some reservations about nationalization early in 1973, when, at a press conference in Benghazi, he said:

> To speak of nationalization is like asking for war in Egypt. Who would buy the oil? When BP was nationalized [in Libya], the Soviets agreed to buy only a very small part of its production from us. Look at Iraq's difficulties . . . It is better to proceed very slowly, to set up some refineries, to train staffs.[6]

However, he proceeded to buy out 51 percent of the Libyan concessions, and nationalized most of the foreign companies working in Libya. There could be only one reason for such action; it was solely political. As Henri Madelin rightly points out:

> Nationalization is impossible while a country has insufficient facilities and skilled workers *but the possibility can serve as a long-term threat or instrument of reprisal* . . .
> Since a 'progressive' government came to power in Libya many moves in the direction of nationalization of the oil companies have been made. The operators most affected are the British and American groups, because of particularly strong reaction of the military regime to the military policy of their countries, which had established important bases in Libya . . .
> *In the final analysis [nationalization] is a political matter and economic rationality is not necessarily applied where political considerations are predominant.*[7]

(Emphasis added)

In May 1973 Colonel Gaddafi delivered an ultimatum to the oil companies to accept 100 percent Libyan control of oil operations. Compensation for the take-over would be based on 'net book value' only, and Libya reserved the right to sell the oil at prevailing market prices. Gaddafi did not even wait for the companies' reaction. On 11 June, at the celebration commemorating the third anniversary of the expulsion of 'United States imperialist forces' from Wheeler air base, he announced that he had expropriated the properties of Nelson Bunker

Hunt to give the United States 'a big hard blow on its cold insolent face'.[8] An official Libyan decree warned the United States 'to end its recklessness and hostility to the Arab nations'.

However, not all nationalizations have been directed single-mindedly against all Western governments. When Iraq nationalized IPC, it offered France the opportunity to buy at cost the 23.75 percent share of the nationalized oil owned by CFP, the same as it had been paying under the concession. This offer was an explicit acknowledgement of 'the just policy pursued by France towards Arab causes and more specifically towards the Palestine cause'.[9]

Even during nationalization by Iraq, there was still scope for rewarding political sympathizers — yet another demonstration that it is impossible to separate politics from Arab oil policies. As pointed out by *The Economist*, the first thing to be understood about 'the oil business is that it is more a political than an economic activity'. That could not be more true anywhere than in the Arab world. The oil crisis and the Palestinian cause have been directly linked since the day the Arabs realized the strategic importance of oil as their primary political tool. Judged from this perspective, nationalization, as other forms of action, is directed towards one political objective — the liberation of Palestine!

Notes

1. *Petroleum Times*, 7 May 1971.
2. Ahmed Zaki Yamani, 'Participation versus Nationalization', a lecture delivered at the *3rd Seminar on the Economics of the Oil Industry*, American University of Beirut, Spring 1969.
3. Aramco is now wholly owned by the Saudi government but the company continues to perform technical tasks. Participation is no longer an issue.
4. *Middle East Economic Digest (MEES)*, 29 May 1970.
5. Ibid., 6 Nov. 1970.
6. Christopher T. Rand, *Making Democracy Safe for Oil: Oilmen and the Islamic East* (Little, Brown & Co., Boston, Mass., 1975), pp. 132-3.
7. Henry Madelin, *Oil and Politics* (Saxon House, D.C. Heath, London, 1974), pp. 73 and 75.
8. *New York Times*, 12 Aug. 1973.
9. *MEES*, 16 June 1972.

PART FOUR
ARAB OIL STRATEGY

13 ARAB OIL STRATEGY

Since the First World War, oil has become an economic commodity of critical strategic importance to the balance of world power. In addition to its vital importance during war-time, oil has come to play a basic role in peace-time economic activity, especially in the post-1945 era.

American strategists realized early on the importance of oil resources in general, but they were particularly interested in Middle East oil and its possible control by American oil companies. Just as it was becoming clear that the Middle East possessed the large oil reserves in the world, American oil companies, with assistance from the US Department of State and Trade, began to take up significant concessions in the region. During the 1920s Standard Oil of New Jersey and Mobil were admitted to Iraq. More concessions were acquired in Kuwait, Bahrain and Saudi Arabia. By the end of the Second World War, United States oil companies had acquired a dominant strategic position in the balance of the world oil industry. With its newly acquired status in world politics, Middle Eastern oil became a tool which the United States used to better its position in Western Europe *vis-à-vis* the Soviet Union and 'international communism'. Secure access to oil supplies and the protection and advancement of American interests in the Middle East were accorded high priority among United States goals in the post-war period, but containment of 'international communism' was the major political issue for United States foreign policy:

In Europe, in the Middle East, in Korea, in the Third World, the United States strove endlessly for the containment of what was thought to be an aggressive and subverting force controlled from Moscow. Through two decades, the Truman Doctrine, the Marshall Plan, NATO, SEATO, CENTO, military assistance pacts with 42 countries, and open and secret warfare from Iceland and Vietnam were all varying manifestations of this one dominating doctrine.[1]

Oil and the Middle East played an important part in the implementation of this policy. In terms of priorities, Western Europe was a vital area for American 'national security', an area of core interest.[2] However, the Middle East was of special interest to post-war American policy, as both the Truman Doctrine and the Marshall Plan were linked

123

to Middle Eastern oil.[3] James A. Forrestal, the leading business voice in the Truman administration, declared in a memorandum to the President: 'Without Middle East oil, the European Recovery Program has a very slim chance of success. The US simply cannot supply that continent and meet the increasing demands here.'[4]

In his book, *Middle East Oil and the Energy Crisis*, Joe Stork observes that:

> the Truman Doctrine, with its provisions for American military aid to Greece and Turkey, was promoted as being essential to the security of Europe and the United States primarily because of Greece and Turkey's geographical position in relation to the Middle East and its oil. All unseemly references to 'oil' and 'natural resources' were expunged from the final text of Truman's message to the Congress in favor of high-sounding abstractions like 'democracy' and 'freedom', but an early draft of the Truman Doctrine contained this contribution from Clark Clifford: 'If, by default, we permit free enterprise to disappear in the other nations of the world, the very existence of our own economy and our own democracy will be gravely threatened . . . This is an area of great national resources which must be accessible to all nations and must not be under the exclusive control or domination of any single nation.'[5]

From a relatively late and modest beginning in 1901,[6] the Middle East had become by the end of the Second World War the largest pool of oil in the world. Moreover, by the early 1970s, oil had become not only the area's most successful commercial commodity, but also the origin of its political strength and diversity.

Arab Oil Strategy in Action

On Saturday 6 October 1973 Egypt and Syria launched full-scale military operations against Israel. Their declared aim was the liberation of Arab lands occupied by Israel during the 1967 war. Within two weeks, the Arabs tightened their grip on oil production, and the repercussions spread rapidly around the world. The war continued for only three weeks before the United Nations arranged a cease-fire, but it brought to light the fact that the Arab–Israeli conflict is but one facet of a complex network of international relations influencing the destiny of the Middle East.

The 'oil war' started even before the outbreak of hostilities between the Arabs and Israel in October. For the first time, all Arab producing countries agreed that some action based on oil had to be taken before the Arab-Israeli dispute exploded, because they feared the explosion would not be limited to the area but would spread all over the world. Preparations for the oil war had begun months in advance, prompted by the absence of a peaceful solution to the Palestinian question, by Israeli intransigence, and by America's obstinate pro-Israel bias, all of which convinced the leaders of the Arab oil-producing countries that recourse to the oil weapon was necessary for the liberation of Arab lands.

The preparations took various forms. On 11 July 1973 Gaddafi nationalized all foreign companies operating in Libya, thereby, as *Newsweek* commented, 'threatening to set off an oil war between the petroleum-rich Arab world and the thirsty industrialized West.' Announcing the nationalization, Gaddafi declared:

It is high time the Americans took a strong slap on their arrogant face. American imperialism has exceeded every limit . . . The American support of our Israeli enemy threatens our security with their aircraft carriers and, from time to time, the Americans threaten our territorial waters.[7]

The following day, President Sadat of Egypt hailed the Libyan decision, calling it 'an opening of the battle with American interests in the Arab area'. He further added that 'this decision is a fundamental part of the formidable conflict and is the first spark triggering the start of the battle'.[8]

Saudi Arabia took a leading position in the oil war effort. Initially attempting to play the role of mediator, albeit unsuccessfully, it soon moved to support escalation of oil warfare. Having a long-standing friendship with the Western powers, King Faisal felt obliged to advise them of the hazards and implications they might face if they continued a one-sided Middle East policy. In April 1973 the King dispatched a diplomatic mission composed of Saudi Oil Minister Sheikh Zaki Yamani and Deputy Oil Minister Prince Saud al-Faisal to Washington with a warning. At the time, Sheikh Yamani publicly linked oil and Israel for the first time. It was politically impossible, Yamani told American officials, for Saudi Arabia to expand production at the desired rate unless the United States changed its policy towards Israel. The Nixon administration ignored the Saudi warnings.

In August, Saudi officials underscored their warning to the West that Arab oil and Arab financial surpluses would be used as tactical instruments in the Arab-Israeli conflict. Prince Abdullah ibn-Abdulazia, Second Deputy Prime Minister and Commander of the National Guard, in an interview with the Lebanese weekly *Al-Hawadis*, declared that the Supreme Oil Council of the Kingdom did not concentrate solely on economic issues but that its purpose included 'the derivation of the maximum benefits from using oil in the service of Arab causes'. Similarly, Prince Saud al-Faisal, Deputy Oil Minister at the time, stated on 31 August in the same weekly that Arab oil and Arab petrodollars would be used as political weapons in the Arab-Israeli conflict. Again, these warnings fell on deaf ears, the Nixon administration choosing to ignore them and even going so far as to dismiss them as frivolous.

Saudi warnings escalated. On 31 August, in an interview with the National Broadcasting Corporation (NBC), and again in September in *Newsweek*, King Faisal himself made it clear that Saudi Arabia would use its oil to political advantage if the United States continued to support Israel's policy of aggression against the Arab world. In the *Newsweek* interview, the King stated:

> Logic requires that our oil production does not exceed the limits that can be absorbed by our economy. Should we decide to exceed the limit in response to the needs of the United States and the West, two conditions must be satisfied. First, the United States and the West must effectively assist the Kingdom of Saudi Arabia in industrializing itself in order to create an alternative source of income to oil, the depletion of which we shall be accelerating by increasing production — and also so that we can absorb the excess income resulting from production at such a level. Secondly, the suitable political atmosphere, hitherto disturbed by the Middle East crisis and Zionist expansionist ambitions, must be present.[9]

A number of American observers warned of the serious nature of the Arab threat and advised a more even-handed policy towards the Middle East. James E. Akins, later to become American ambassador to Saudi Arabia, warned that the United States should pay heed to King Faisal's announced intention of refusing to increase oil exports. Unfortunately, American policy remained unchanged, and when war broke out, Saudi Arabia led the other Arab oil-producing countries in an effort to capitalize on the political impact of oil power. As Ruth S. Knowles wrote:

Unfortunately, there were too few on the American political scene who recognized, even at this late date, the importance of improving US policy towards the Arab countries in a more 'even-handed' direction. Nor did they read as carefully as the Arabs already had done the US energy reports, which were now pouring out, showing the magnitude of our future dependence on Arab oil.[10]

The Kuwait Conference

On 17 October 1973 OAPEC members met in Kuwait, and decided to cut back oil production and exports by 5 percent from the September levels. The cutback was to be increased by an additional 5 percent each month until all Israeli forces were withdrawn from the occupied Arab lands. OAPEC also decided to embargo all exports from Arab countries to the United States and Holland.

At the conference, discussions of the use of Arab oil power led in two directions, one advocating the reduction of oil production, and the other advancing the nationalization of foreign, particularly American, oil interests in the Arab world. The first view was held by Egypt's Oil Minister Ahmed Helal, who said in his opening speech to the conference:

President Sadat's message to you is that the situation cannot sustain any failure or it will result in a political setback for the Arab position which might lead to undesirable consequences for our battle's outcome. It is important to reach a unanimous decision . . .

It is necessary, at least at this stage, to preserve Arab financial capabilities. As such, a decision to implement a complete cutback [of oil production], or nationalization might give the other parties [Western powers] a cause to freeze Arab financial assets.

The Oil Minister expressed Egypt's view that it was important for the Arabs not to alienate the international community, particularly the industrialized countries, from the Arab world at a time when their support was most needed. He also stressed that it was of vital importance to convince the Western industrialized countries to exert pressure on the United States to adjust its one-sided policy, and to convince the United States, in turn, to pressure Israel to accept a political settlement of the conflict. Helal concluded his speech by saying:

Our aim is to reach a decision which could be implemented immediately, [a decision] which would have a sensible impact that escalates by the passing of time.

We want a gradual reduction in the rate of Arab [oil] production which would, in its first stage, be directed towards the United States and expand to reach other areas of the world. We want at the same time to invite the countries which proclaimed neutrality toward the prevailing struggle to change their positions toward our cause.[11]

The other view expressed at the conference called for the complete nationalization of foreign, particularly American, oil companies operating in the Arab world. The main advocates of this position were Iraq and Libya. As expressed by Dr Sa'doun Hammadi, the Iraqi Oil Minister, the Arab producing countries should:

(1) nationalize all American oil interests in the Arab world;
(2) withdraw all Arab financial reserves from the United States;
(3) sever diplomatic relations with the United States.

The Iraqi delegation also insisted on punishing the United States, demanding that at a minimum the conference should boycott oil exports to America, or the Iraqi delegation would dissociate itself from the conference resolutions and not attend the final session.

The Libyan Oil Minister, Ezzedin al-Mabruk, supported the Iraqi position and put forward the following suggestions:

(1) to deny the United States crude oil and natural gas and other products immediately;
(2) to reduce Arab oil production by the amount previously exported to the United States and to persuade non-Arab oil-producing countries (Iran, Nigeria and Venezuela) not to supply the United States with the needed oil;
(3) to withdraw Arab financial reserves from the United States and to direct Arab financial experts to find a way of using these reserves in Europe to sabotage the United States' world financial position.

Following intense deliberations, however, the Arab oil ministers unanimously agreed on the 'gradual reduction' approach. Production cuts were directed against all countries that were pro-Israel. The main target, of course, was the United States, which had by then started to 'pour' weapons into Israel. The official declaration by OAPEC stated:

The Arab oil ministers meeting on 17 October in the city of Kuwait have decided to begin immediately the reduction of production in every Arab oil-producing country by no less than 5 percent of the production for the month of September. The same procedure will be applied every month and production will be reduced by the same percentage of the previous month's production until the Israeli forces are completely evacuated from all the Arab territories occupied in the June 1967 war, and the legitimate rights of the Palestinian people are restored.[12]

On 4 November, at their second meeting in Kuwait, the OAPEC members decided to increase the production cutback by 25 percent as a means of increasing pressure to achieve an Israeli withdrawal from Arab territories. However, countries considered 'friendly' to the Arab cause were to be given special treatment, enabling them to maintain their current level of Arab oil consumption. No particular countries were labeled 'friendly', but it was assumed that they were Britain, France, Spain, India and Pakistan. The United States then announced a grant of $2.2 billion in military aid to Israel, which led the Arab states to embargo all exports to the United States. The cutback in Arab oil production, together with the American embargo, created near panic buying, forcing market prices of crude oil to skyrocket.

The Instruments of Arab Strategy

The oil war was fought on three fronts: reduced oil production; embargoes, principally against the United States and the Netherlands; and steep rises in oil prices. A fourth 'front' could also be added − the nationalization of foreign oil companies. This last measure, however, was considered unilaterally by each Arab country, whereas this chapter concentrates on the collective measures adopted unanimously by all Arab producers after the outbreak of the Ramadan war.

The Cutback of Oil Production

For the first time in Arab oil history, the events surrounding the Ramadan war found the Arab states in near unanimous support of Arab political objectives. Under the leadership of Saudi Arabia, they decided to use the 'oil weapon' on behalf of the Arab cause. Of the ten members of OAPEC, only Iraq refused to abide by the Kuwait decision. Iraq favored nationalization of American and Dutch oil interests in

line with its action of 6 June 1973.[13] Saudi Arabia and the majority
of OAPEC members believed that a gradual cutback in oil production
would be more effective, for the following reasons:

1. Since the Arab armies were judged to be militarily strong, participants in the Kuwait meeting were encouraged to adopt a moderate
plan of action.
2. The Saudi government still hoped to persuade the United States
to modify its policy of active support for Israel, and keeping the
boycott provision vague would avoid a 'point of no return' show-
down with the United States.
3. Saudi Arabia thought of this strategy as an initial move which
would give the Arabs enough flexibility to curtail production further
if the need arose.
4. Because the war with Israel might continue for a long time, the
'gradual reduction' approach was more expedient than an immediate
embargo or nationalization, and would allow the Arab oil producers
enough revenue to finance the military activities of the Arab states.
5. Such an approach would mean that the 'neutral' and 'friendly'
countries would not suffer too much, since the oil weapon was intended to hurt only those countries which supported Israeli expansionist policies.

The initial 5 percent oil cutback led to a hardening of the terms by
OAPEC members. The 5 percent cutback increased and, within a few
days, reached over 25 percent of the September production level. Table
13.1 shows the relation of the cutback to the September 1973 levels of
output.

The Oil Embargo

In addition to the cutback in oil production, a second 'weapon' was
used in the oil war. A selective embargo was imposed upon countries
which openly supported Israeli policy, with the United States at the top
of the list. Other embargoed countries were the Netherlands, Portugal,
South Africa and Rhodesia.

The oil embargo was decided by the Arab oil-producing countries on
19 October 1973, when President Nixon asked the United States
Congress to approve $2.2 billion in emergency military aid to Israel.
As explained by Nixon, Israel needed military assistance 'to maintain
a balance of forces and thus achieve stability'. The Arab countries
were outraged at Nixon's declaration.

Table 13.1: Arab Oil Cutbacks in November 1973 (hundred barrels per day)

	September 1973 production	Production after cutbacks	Change in volume	Percentage of cutbacks
Saudi Arabia	85,499a	58,494	27,005	31.5
Kuwait	35,063	24,788	10,275	29.0
Iraq	21,115	19,560	1,555	7.3
Abu Dhabi	13,980	10,485	3,495	25.0
Qatar	6,086	4,584	1,502	25.0
Libya	22,861	17,146	5,715	25.0
Algeria	10,500	7,875	2,625	25.0
Other countriesb	11,000	8,250	2,750	25.0
Total	206,104	151,182	54,922	26.7

Notes: a. Production figures for Kuwait and Saudi Arabia include volumes produced in the Neutral Zone, which are attributed equally to both countries.
b. Egypt, Syria, Dubai and Oman.
Source: *Arab Oil and Gas Journal* (16 Dec. 1973), reproduced in Assad Abdul-Rahman (ed.), *The Fourth Arab-Israeli War*, Palestine Books, no. 59 (Palestine Liberation Organization (PLO), Beirut, 1974), p. 491.

The United Arab Emirates (UAE) was the first Arab producing state to impose a total embargo on the United States. As stated by Sheikh Mana Saeed al-Otaiba, the UAE's Oil Minister:

We see that in this critical stage in which the [Arab] nation is waging an honorable battle against the Zionist enemy and [against] the powers that support it with the means of survival, my Government has no choice but to declare its willingness to put all its capabilities at the service of the battle . . . We believe that the use of the oil weapon would give its fruits if all the Arabs used this weapon.

Accordingly, Abu Dhabi's government must define its relations with all nations in the light of their stand on the battle. Abu Dhabi's government has decided to cut oil supplies from the United States until it changes its aggressive stand against the Arab nation.

This measure will be extended to any country which allows the same aggressive attitude toward the Arab nation in its battle of destiny.[14]

On 20 October the Saudi government announced that 'in view of the increase of American military aid to Israel, the Kingdom of Saudi Arabia has decided to halt all oil exports to the United States of America for taking such a position'. The Saudi action was not hastily taken: King Faisal tried all possible means of diplomatic communication with Washington. Even when news of the substantial American aid to Israel was announced, the King preferred to take a low-key position. This explains why, on 18 October, Saudi Arabia limited its action to a cut in exports of 2 percent (instead of the mandatory minimum of 5 percent) and why it refrained from immediately proclaiming an embargo. However, at the same time, the Saudi government urged the United States:

to change its present stand toward the waging of war between the Arab nations and Israel and stop its military assistance to Israel. If Saudi efforts to compel the US to take a neutral stand should not bring about, immediately, a sensible result, Saudi Arabia will halt its oil exports to America.[15]

By 22 October all Arab producers had followed the Saudi action and embargoed shipments to the United States. Some added the Netherlands to the list as well. As Joseph S. Szyliowicz pointed out:

The embargo of October dramatically highlighted the vulnerability of the industrialized nations of the West. As the weeks passed, however, the price dimension came increasingly into prominence, as economists warned of the consequences of a new oil structure for national economies and the international monetary system.[16]

The Price Explosion

The third and most serious consequence of the oil war was the dramatic and unprecedented increase in the price of crude oil. On 16 October 1973 all Arab countries decided to raise crude oil prices by 70 percent. This decision was a direct consequence of a scheduled meeting between the oil companies and the Gulf producing countries to negotiate increased oil prices. Negotiations began in the second week of October, at which time the companies asked for a two-week recess to consult with major consumer governments. At this point, the OPEC countries' demands were for a rise in posted prices of between 35 and 50 percent. Refusing the postponement, OPEC scheduled a meeting for 16 October in Kuwait, and it was then that the decision to increase oil prices by 70 percent was taken.

The motive behind the decision became clear only after reviewing the statement made by the representatives of the Gulf states on 16 October:

On the basis of the decision taken in Vienna, on 12 October, the ministerial committee met in Kuwait on October 16 and decided:
1. On the basis of OPEC's decision number 90, and of the actual practice of other members in the organization — Venezuela, Indonesia and Algeria — it is decided to fix and publish the posted prices for crudes in the Gulf.
2. The new posted prices are based on the realistic market prices in the Gulf and in other regions and are designed to reflect differences in the API degree and in the geographic location.
3. Starting from 16 October, realized prices will determine the level of posted prices in such a way that the relative correspondence between the two prices — this correspondence which held in 1971 before the Tehran agreement — is not altered. The raising or lowering of posted prices will take place depending on whether the market price exceeds or falls short by 1% of the price adopted to determine the posted prices.
4. The market price for Arabian Light corresponding to the new

posted prices amounts to $3.65 per barrel. The prices of other crudes are determined on the basis of this price which exceeds the latest sale prices for this crude by only 17%. Accordingly, the posted prices of all crudes will rise in the same proportion.

5. The premium on low sulfur crude is determined unilaterally by individual OPEC member countries on the basis of market trends. [Abu Dhabi and then Qatar specified this premium; Iran also decided on a small premium for Iranian Light.]

6. The Geneva agreements remain in force.

7. The new arrangements and prices become effective starting 16 October 1973.

8. If the oil companies refuse to buy the crudes on the basis of the new arrangements, producing countries will provide each buyer with the various crudes at prices calculated according to the price of Arabian Light equivalent to $3.65 per barrel, Ras Tannura port delivery.

This was the first time that the oil-producing governments were able collectively to set their prices upon world markets. This unprecedented move successfully increased the price from $1.77 to $3.05 and later to $7.00 per barrel.

Since the early 1970s, the producing nations had strongly advocated price hikes. Previously, OPEC price increases had never exceeded 35-50 percent; a 400 percent increase within a matter of months was totally unexpected (see Table 13.2).

In this way, the oil-producing countries were finally able to break the chains of servitude to the major oil companies. The companies had no choice but to accept the decisions of the oil producers and accede to the price increases. The companies were reluctant, but they acquiesced because supplies fell far short of demand as a direct result of the Arab oil cutback decisions of 17 October 1973.

The situation prevailing before the oil price revolution of 1973-4 was probably not stable; in any event, the principal consumers of energy would sooner or later have concluded that, even if they could continue safely to disregard the oil producers' preferences extraction would eventually have to be regulated in accordance with their relative preferences as between immediate and future needs, and prices would have to rise. It was the abruptness and magnitude of the price change in 1973-4 that roused passions. Until the early 1970s, price control had been in the hands of the Majors. When OPEC was established in 1960, their only achievement was the freezing of oil prices at their 1960 level.

Table 13.2: Arabian Gulf Crude Oil Prices[a] (US dollars per barrel)

	1971 (before Tehran) $	1971 (after Tehran) $	Percentage change	February 1972 (after Geneva) $	Percentage change, 1971-2	January 1973 (pre-October oil war) $	January 1974 (post-October oil war) $	Percentage change, 1973-4
1. Posted price	1.800	2.285	+ 27.0	2.479	+ 8.49	2.591	11.651	+ 350.0
2. Royalty[b] (12.5% of 1)	0.225	0.285	–	0.310	–	0.324	1.456	–
3. Production cost[c]	0.100	0.100	–	0.100	–	0.100	0.100	–
4. Tax (55% of 1 + 3)[d]	0.770	1.045	–	1.138	–	1.192	5.552	–
5. Government revenue (2 + 4)	0.995	1.330	+ 33.6	1.448	+ 8.87	1.516	7.008	+ 362.0

Notes: a. Prices shown are for Saudi Arabia Light crude oil 34° API (American Petroleum Institute) gravity. Saudi Light is used as a standard for Arabian Gulf crude.
b. The Saudi royalty was fixed at 12.5 percent of the posted price for the 1973 and 1974 rates. However, it increased to 20 percent in 1975.
c. Production costs are estimated at 10 percent per barrel.
d. The Saudi tax was fixed at 55 percent of the profit for tax purposes for 1973 and 1974. In 1975 it increased to 85 percent.
Source: Compiled from *International Economic Reports of the President*, Feb. 1974 and March 1975; also Jean-Marie Chevalier, *The New Oil Stakes* (Penguin Books, London, 1975), p. 61.

As Owais Succari stated:

> Petroleum prices have a primordial importance for the producing as
> well as the consuming countries which, together with the petroleum
> countries, form the international petroleum market. Paradoxically,
> both of the first two parties did not have, until very recent years,
> a word to say on the question of pricing. Petroleum prices were
> looked upon as 'administered prices', that is, they were largely con-
> trolled and directed by a small number of firms, the Majors, who
> acted as price makers or price leaders.[17]

At their second meeting, on 23 December, OPEC announced a 130
percent increase in crude oil prices, raising the posted price from the
1 November figure of $5.17 per barrel to $11.65, beginning 1 January
1974. This new price hike increased government's share to $7 per
barrel. In announcing the increase, the Shah of Iran declared:

> The industrialized world will have to realize that the end of the era
> of their terrific progress and even more terrific income and wealth
> based on cheap oil is finished. They must find alternative sources of
> energy. Eventually they'll have to tighten their belts.[18]

However, Libya, as usual, followed its own path. Although it went
along with the OPEC decisions, Libya announced a world record posted
price of $18.76 per barrel for its oil, effective from 1 January 1974.
Nigeria also boosted the posted price of its crude to $14.67 per barrel.

The participation of non-Arab OPEC members in the price escalation
notwithstanding, there was a common assumption in the West that the
price increases were the outcome of temporary circumstances which
had nothing to do with the world oil market or the supply and demand
mechanism. The Arab oil boycott reduced production and the embargo
contradicted every existing economic rule. As Professor Walter J. Levy
pointed out, 'economic factors did not matter in deciding oil prices'.
Although the cutback in oil exports was exclusively a political decision,
oil price rises were a mixture of both political and economic factors.
Demands for higher oil prices were pressed before the October 1973
war, but the war made the increases possible. Without the interruption
of oil supplies in October–December 1973, oil prices would surely not
have increased so fast or to such heights.

Notes

1. Carl Solberg, *Oil Power* (Mason/Charter, New York, 1976), p. 197.

2. As defined by Professor Fred Warner Neal, a 'core interest' is one which, when it is threatened, is considered so vital that the threat is regarded as a danger to the very existence of the nation itself. See F.W. Neal, 'The Theory of Core Interest and US-Soviet Cold War Rivalry', a paper presented at the Second Annual Meeting of the International Studies Association, 29 March 1961, p. 3.

3. Solberg, *Oil Power*, p. 176.

4. 'Memorandum to the President', in *Forrestal Papers*, (Princeton University Press, Princeton, 1948), quoted in ibid., p. 179.

5. Joe Stork, *Middle East Oil and the Energy Crisis* (Monthly Review Press, New York, 1975), p. 40.

6. The first oil concession was granted to William D'Arcy in 1901 by the Shah of Persia (Iran). The first discovery was made in 1908 at Masjid-i-Sulaiman.

7. *Guardian*, 17 June 1973.

8. 'The Arab Oil Squeeze', *Newsweek*, 17 Sept. 1973, p. 15.

9. Ibid., p. 16.

10. Ruth S. Knowles, *America's Oil Famine* (University of Oklahoma Press, Norman, Oklahoma, 1980), p. 108.

11. Mohamed Harb, *Oil War: The Secret Minutes of the Arab Oil Ministers Meeting* (General Printing Company, Cairo, 1974) (in Arabic), pp. 116-17.

12. Press Release of the OAPEC Ministers Meeting, Oct. 1973.

13. On this date, Iraq nationalized American and Dutch interests in the Basra Oil Company.

14. Harb, *Oil War*, p. 133.

15. Abdulaziz H. al-Sowayegh, 'Saudi Oil Policy During King Faisal's Era' in Willard A. Beling (ed.), *King Faisal and the Modernization of Saudi Arabia* (Croom Helm, London, 1980), p. 212.

16. Joseph E. Szyliowicz and Bard E. O'Neill (eds.), *The Energy Crisis and US Foreign Policy* (Praeger, New York, 1975), p. 185.

17. Owais R. Succari, *International Petroleum Market: Policy Confrontations of the Common Market and the Arab Countries* (Université Catholique de Louvain, 1968), p. 12.

18. *New York Times*, 26 Dec. 1973.

14 THE 1973 RAMADAN WAR'S IMPACT ON THE INDUSTRIALIZED COUNTRIES

The oil war came as a shock to the Western industrialized countries, which had never seriously considered the prospect that one day they might face a major 'energy' problem. However, within a short period of time, the potency of the oil weapon became obvious. Arab oil-exporting countries succeeded in disrupting the life-style of every major industrial power, caused fissures in the Atlantic Alliance, precipitated an upset in the international monetary situation, and prompted the United States to begin an intensive search for a peace settlement in the Middle East. As Ruth S. Knowles observed:

> Not since the Trojan horse and the atomic bomb had a new war weapon been devised that was as devastatingly effective as the Arabs' use of their newly acquired oil power . . . Although oil power for the Arabs had suddenly become a means to achieve a devout, dedicated goal in their quarter century struggle against Israel and Zionism, the success with which they wielded their new weapon had a heady effect . . . The oil weapon was double-edged — political and economic. The Arabs slashed with both sides.[1]

The West was for the first time to feel the pinch of the Arab oil embargo and to reconsider the importance of oil to its civilization. Although oil continues to be the lifeline of Western industrial complexes, the extent of their dependence on abundant, cheap supplies of oil was driven home as never before.

Economic and environmental problems associated with alternative energy sources, the finite and depletable nature of oil resources, and the concentration of oil reserves in Arab lands together translated the West's increasing energy demands into greater dependence on Arab producers. This dependence, and hence their vulnerability, always worried the West. But the effective Arab resort to the oil weapon and the certainty of its continued use as long as the Arab–Israeli conflict remained unresolved brought matters to a head in late 1973. Since then, any new explosion in the Middle East has been feared as almost certain to lead to new embargoes, oil shortages and higher oil prices.

The United States was the main target of the Arab oil embargo.

American support for Israel triggered the Arabs' use of oil as a political weapon. Although the United States had imported oil for at least 50 years, only in the last decade had domestic demand outstripped production capacity. The result was a dramatic increase in imports. In his testimony before the Senate Foreign Relations Committee on 31 May 1973 William E. Simon, then Deputy Secretary of the Treasury, testified that domestic demand for oil had increased from 15.1 million b/d in 1971 to approximately 18 million b/d in 1973 and would jump to about 25 million b/d in 1980.

According to reliable estimates, United States imports of Arab oil (crude and refined) averaged 2 million b/d in August and September of 1973. This quantity of crude and refined products equaled approximately 13 percent of American consumption and about 32 percent of total United States imports of crude and refined products.

Until the mid-1960s domestic production of oil in the United States had exceeded oil imports. In 1972 this situation changed drastically. Texas and Louisiana (the two major oil-producing states) de-regulated oil productions. Thus in 1973, the United States produced 9.2 million b/d, whereas it consumed 17.2 million b/d, which necessitated importing over 46.5 percent of its daily consumption. Canada contributed over 17.8 percent of this import gap, while OPEC and Mexico accounted for the remaining 82.2 percent. By 1980 American oil imports had decreased to 42 percent of domestic availability, but imports of Arab oil increased from 18 percent of total oil imports in 1973 to over 47.3 percent in 1980. This growing dependence on Arab oil led to a study which revealed the following trends:

(1) Middle East oil is likely to remain the most abundant and the cheapest to produce in the world until the next century.
(2) The Soviet bloc is likely to increase its imports of oil and is expected to become a net importer of oil of over 100 million tons per year in the 1980s (after being a net exporter in previous decades).
(3) Oil discoveries in the United States have fallen, with a consequent decrease in the proven reserves to a production ratio below the safe level of ten.
(4) Oil supplies from Venezuela, Canada and Mexico, which have been classified as secure sources, are likely to fall, and Canada has already introduced a quota on oil exports to the United States.
(5) Oil reserves in the North Sea are limited and costly to extract, and likely to be sufficient only for British requirements.
(6) Nuclear energy is costly to produce and still unsafe. (To be

economic, nuclear power stations must be located near cities, but the closer to cities, the greater the environmental danger.)

(7) Oil shales and tar-sands are abundant in the United States. However, oil from such deposits is costly to produce and its production erodes substantial areas of the environment.

(8) The growing oil revenues of the oil-producing countries are likely to be recycled into the United States; and Japan, West Germany and other European competitors are likely to suffer relative to America's improved competitive industrial position.

(9) The United States is likely to be compelled to relinquish its quotas on oil imports, and consequently its imports from the Middle East are likely to rise substantially.[2]

The identification of these trends had an immediate impact on the United States energy policies. The vulnerability of the United States to interruptions in the flow of Arab oil was the core element in the shaping of these policies, as the United States prepared to mitigate this impending peril.

For the first time in 1973, the oil weapon represented a significant challenge to Western countries' attitudes towards the Arab-Israeli dispute. The West European countries and Japan were hardest hit by the oil war, and from the Arabs' point of view, their response was quick and 'positive'. The European Economic Community (EEC) and Japan hastened to reassert their good intentions towards the Arabs and disavow any association with Israel or with United States' support for the Israelis.

The schism that was developing between the United States and Western Europe as a result of these events clearly widened on 6 November 1973, when representatives of the EEC adopted a joint statement. The nine governments, including Holland, strongly urged that both sides in the Middle East conflict return immediately to the positions they had occupied on 22 October, in accordance with Resolutions 339 and 340 of the UN Security Council, and declared that a peace agreement should be based on the following points:

(1) the inadmissability of the acquisition of territory by force;

(2) the need for Israel to end the territorial occupation which it had maintained since the conflict of 1967;

(3) respect for the sovereignty, territorial integrity and independence of all states in the area, and for their right to live in peace within secure and recognized boundaries;

(4) recognition that, in the establishment of a just and lasting peace, account must be taken of the legitimate rights of the Palestinians.

The Ramadan war also proved beyond a shadow of a doubt Japan's basic political and economic vulnerability. When it realized that its political 'neutrality' was an affront to the Arabs, Japan followed Europe's example, and on 22 November 1973 the Japanese Cabinet announced that it might reconsider its policy towards Israel. Then, on 23 December, Japan appealed to Israel to withdraw to the 22 October cease-fire line as a first step towards total withdrawal from occupied Arab territories. Japan's options were limited; it could either follow the United States, which preferred that Japan remain non-committal to the Arabs, and tolerate the embargo for a few months but without any assurance of emergency supplies; or Japan, which had an oil stockpile for 59 days, could ignore American pressure and negotiate with the Arabs at the risk of further deterioration in the American alliance. Japan opted for the latter alternative – a subtle process of de-Americanization.

This trend was clearly in evidence even before the October crisis for, in May 1973, the Japanese Industry Minister, Nakasone (currently Prime Minister), declared in a press statement on his return from a Middle Eastern tour:

I have become strongly aware of the need to approach Middle East oil not simply as tradeable merchandise but something more deeply politically involved. Oil is a critical resource for Japan and dealings in oil cannot be handled by individual Japanese enterprises or traders alone without the support of the Japanese Government and its people. The Japanese Government will involve itself in strong and continuous petroleum diplomacy in the future.

The international oil situation is in a period of transition with producing nations seeking partners among consuming nations for long-range oil contracts. Establishment of a co-operative relationship between a group of the world's largest oil-producing nations and Japan, as one of the world's largest consuming nations, will have an important influence over the international oil scene.[3]

The Washington Energy Conference

Frustrated by European oil 'madness', the Americans called a consumer conference in Washington on 11 February 1974 to work out a joint action program. Invitations were sent to the member countries of the EEC, Canada, Norway and Japan, plus the Secretary-General of the Organization for Economic Co-operation and Development (OECD), to which all these countries belong.

In an interview in *Le Monde*, Algerian President Boumedienne called the Washington conference a 'plan designed to prevent contacts between the oil-producing and -consuming countries'. He further warned that, 'if the Europeans yield before the American "big stick", they will once again return to the sidelines of history'.[4]

Secretary of State Henry Kissinger, on the other hand, stressed that the conference should not be taken as a confrontation between consumers and producers. On 6 February he said that, after the conference 'there will be a conference first between the nations that are coming next week and the consuming less developed countries. And then very shortly after that, a meeting between all the consumers and the producing countries.' He further stressed that:

the USA has called this conference for one central purpose — to move urgently to resolve the energy problem on the basis of co-operation among all nations. Failure to do so would threaten the world with a vicious cycle of competition, atrocity, rivalry and depression such as led to the collapse of world order in the thirties.[5]

The 11 February meeting was attended by representatives from all 13 of the major oil-importing, industrialized countries. It was scheduled to end the following day but was delayed by a dispute between France and other EEC members over their response to the United States proposal for a joint effort to confront energy issues.

The contending positions taken by the United States and France advanced two alternative strategies for individual consuming countries to follow in order to achieve price security for imported oil supplies. The American position called for a defensive consumer organization, encompassing protective commercial and financial arrangements, with the United States taking the leadership role in its creation. Collective negotiation with the producer 'cartel' was not ruled out, but it was to be postponed until the consumers had had time to build their strength in organization, consumption control and perhaps even in the development of alternative resources.

On the other hand, the French argued that such a consumer organization should be rejected since it was designed to be headed by the

Americans. As French Foreign Minister Jobert put it, 'Energy matters had been a "pretext", and the conference's real purpose had been the United States "political desire" to limit Europe's and Japan's freedom of action and keep them under American control.'[6] The French argued that the consumer organization advocated by the Americans would antagonize the oil producers and be counter-productive. The French also urged that negotiations with the oil-producing countries be pursued immediately.

With the exception of France, the conference agreed on the need for a comprehensive action program that would co-ordinate national policies in the following areas: (i) conserving energy and restraining demand; (ii) allocating oil supplies during an emergency or severe shortage; (iii) supporting national development of additional energy sources to enable the diversification of supplies; and (iv) accelerating energy research and development programs through international co-operative efforts.[7] Despite French opposition, Kissinger called the conference a success and said that it was a major step toward 'dealing with world problems co-operatively'.[8]

In reality, the conference was dead before it was born. In its rejection of bilateralism, the Washington accord bypassed international companies which were negotiating package deals of industrial exports and investments in exchange for oil supply agreements. As soon as the conference was over, many oil technology exchanges were concluded between European countries and OPEC members. However, the big blow came in early March 1974. The EEC announced plans for another high-level conference on long-term economic, technical and cultural co-operation between Europe and the Arab world. No mention was made of the political question of oil. The State Department officially objected to the EEC plan, complaining that there had been no consultation with the United States. As United States government spokesmen have testified, the United States continued 'to question the advisability and viability of bilateral arrangements to tie up specified amounts of oil in exchange for specific goods and projects . . . These will tend to sustain higher prices.'[9] However, on 8 June 1974 the United States signed a bilateral agreement with Saudi Arabia; the two governments planned to hold negotiations 'to expand and give more concrete expression to their relationship'.[10]

Accordingly, the West European countries argued that the United States was inconsistent, since its declared policy was opposed to bilateral supply agreements. Throughout the discussions with the Saudis, however, the United States denied that its motives were to obtain specific supply commitments:

> We are not engaged in discussion with the Saudis to gain a preferred position with respect to the purchase of Saudi Arabian oil . . . The United States will continue to press for multilateral solutions to world petroleum problems.[11]

As far as the economic aspects of the oil war are concerned, the most obvious consequence of the OAPEC–OPEC action was the redistribution of wealth. While the industrialized countries sank into recession – although oil was not the only contributing factor – oil-producing countries were suddenly gorged with money. With the transfer of wealth went a transfer of power. The 'oil weapon' challenged the balance of power between Israel and the Arab states. However, it would be misleading to measure such a shift in power solely in economic terms.

Whatever the economic consequences of the oil crisis for the industrialized countries, economic hardship can be met effectively, given time for adjustments in production and consumption patterns. But the political impact was the most dramatic result, a 'turning-point in the history of contemporary Europe'.[12] As described by Walter Laqueur:

> The oil restrictions will not last forever and in the long run they will have a salutary effect, forcing the industrialized world to face a problem neglected for too long . . . The real victim of the crisis in Europe was not the growth rate and prosperity, it was the myth of European power and unity. It suddenly appeared that Western Europe with 40 percent of the world's gold resources, 30 percent of its foreign trade, and 20 percent of its population, counted for precisely nothing in terms of political power.[13]

Thus the October 1973 conflict and the oil war triggered a crisis more far-reaching than anyone could have predicted. While the European states advanced bilateralism, the United States insisted that it would never be 'blackmailed' and urged a collective approach to deal with the oil crisis. As a superpower, it was not easy for the United States to change its position. But it, too, resorted to bilateral negotiations in the face of Arab oil.

Notes

1. Ruth S. Knowles, *America's Oil Famine* (University of Oklahoma, Norman, Oklahoma, 1980), p. 107.

2. Marwan Iskander, *Al-Da'm al-Nafti al-Arabi: Min Mu'tamar al-Khartoum 1970 ila Mu'tamar al-Kuwait 1973* (Arab Oil Support from the Conference of Khartoum 1970 to the Conference of Kuwait 1973) (in Arabic) (Middle East Economic Consultancy, Beirut, 1973), pp. 83-4.

3. Makota Momoi, 'The Energy Problems and Alliance Systems: Japan', *Adelphi Papers*, no. 115 (International Institute for Strategic Studies, London, 1975), p. 26.

4. *Middle East Economic Survey*, 8 Feb. 1974.

5. Quoted in S. Manoharan, *The Oil Crisis: End of an Era* (S. Chaud & Co. Ltd., New Delhi, 1974), p. 98.

6. John Maddox, *Beyond the Energy Crisis: A Global Perspective* (McGraw-Hill, New York, 1975), p. 131.

7. United States Departments of State, Washington Energy Conference Communiqué, 17 Dec., Rev. 2 (13 Feb. 1974).

8. 'The Middle East, US Policy, Israel, Oil and the Arabs', 2nd edn, *Congressional Quarterly*, (Oct. 1975), p. 35.

9. 'US-Saudi Agreement Seen as Good Omen for Oil', *Oil and Gas Journal*, 15 April 1974.

10. *Washington Post*, 6 April 1974. (Text of Joint Statement on Co-operation, 8 June, *Department of State Bulletin,* 1 June 1974.)

11. 'US-Saudi Agreement', *Oil and Gas Journal,* 15 April 1974.

12. Edward Friedland et al., *The Great Detente Disaster* (Basic Books, New York, 1975), p. 3.

13. Walter Laqueur, *Confrontation: The Middle East and World Politics* (Quadrangle Books, New York, 1974), pp. 267-8.

15 OIL AND ARAB-AMERICAN RELATIONS

The importance of the Middle East, and particularly the Arab world, to the United States cannot be exaggerated. American interests in the area are dominated by three elements – oil, security and the Arab-Israeli conflict.

The dependence of the United States on Arab oil has already been documented. American interest in maintaining a steady flow of this vital commodity assumed critical importance in the 1970s: US oil imports from the area escalated, and the Soviet Union and its allies verged on becoming net importers of oil. The proximity of the oil-wells to the Soviet Union and their relative distance from the United States underscores the importance to America of geopolitical alliances in the area. At the juncture of three continents, the Arab world flanks Europe, Africa and Asia. For the United States, it would inconceivable for this area to fall under Soviet control or even influence.

Oil alone, one would think, is sufficient reason for the United States to placate and accommodate Arab concerns. Geopolitical considerations add another critical dimension of American interest. Nonetheless, the United States commitment to Israel has dominated every other consideration, including issues of human rights.

The overwhelming American support for Israel created a rift in Arab-American relations and was the major cause of King Faisal's decision to politicize Saudi oil during the Ramadan war. Ghazi al-Gosaibi, Saudi Minister of Industry and Electricity, said in a speech delivered to the American Council on Foreign Relations in New York that:

> This uncritical support [for Israel] prevented Saudi-American friendship from reaching all its potentialities; in 1967 and 1973 it strained relations almost to the breaking point. Had it not been for this single factor, American-Saudi friendship would have been a textbook case of super-power-small-power partnership.[1]

Well aware of the 'special' American-Israeli relationship, King Faisal worked to 'neutralize' the United States' position towards the Arab-Israeli issue. Following the June 1967 war, President Nasser was convinced that there could be no solution to the Arab-Israeli conflict

146

without American participation, and neutralization of the United States towards the Middle East conflict became the Arabs' major theme. Mohamed Heikal, a major advocate of this policy, argued that neutralization of the United States was a prerequisite to the battle with Israel, which he felt was inevitable. He stated that:

> while we [the Arabs] have to accept the fact that the interests of America and Israel are closely linked, we should always strive to ensure that they do not become completely identified with each other. We must work to preserve a gap between the interests and policies of the two countries, and in the gap we should find room to maneuver and to bring pressure on Israel. The assets on which we could count for the achievement of these ends include our ability to wage a limited war, our close relations with the Soviet Union (providing they are kept informed of our intentions), the oil weapon, and the solidarity of other Arab countries.[2]

Egypt was eager to find a way out of the crisis following the 1967 war, and was fully convinced that there would never be a settlement of the lengthy Arab-Israeli conflict unless Washington were to take part in its formulation. But the Egyptian government did not have much confidence that President Johnson was committed to effecting a just settlement, given the biased role he had played in the June war on behalf of Israel. Immediately after the election of Richard Nixon as the new President, Nasser sent him a congratulatory telegram for the purpose of thawing relations between Cairo and Washington. In 1969 Dr Mahmoud Fawzy was sent to Washington as a special envoy, to attend President Eisenhower's funeral. This was followed by Sisco's visit to Cairo in April 1970, then by the Rogers' initiative, and by Elliot Richardson's attendance at President Nasser's funeral. These contacts continued, despite the formal severance of diplomatic relations between the two countries after the June war of 1967.

Anwar Sadat, who succeeded Nasser as President in September 1970, also recognized the importance of the American role. He perceived the situation as one where 'all the cards in this game are in the hands of the United States . . . because they provide Israel with everything and they are the only [ones] who can exert pressure on Israel'. Sadat moved to mend relations with other Arab states, particularly the 'front-line' states of Syria, Lebanon and Jordan. The oil-producing states were even more important to Sadat than the rest of the Arab states. Saudi Arabia was accorded top priority on Sadat's list. The new Egyptian

President was eager to open a new chapter in Saudi–Egyptian relations, strained after Nasser's Yemen episode. After Sadat's consultation with King Faisal, who had established himself as an undisputed influence in Arab policy, Cairo and Riyadh came to play a dominant role in the Arab world. When the Ramadan war erupted, it was these two countries, together with Syria, which were the corner-stone of Arab unity during and immediately following the war.

Sadat also planned to transform Egypt into a private enterprise economy, thus allying itself with the United States rather than with the Soviet Union. Because diplomatic relations with the United States had been severed after the June 1967 war, Sadat's first step was to lead the Americans towards an understanding of the Egyptian position.

The United States was no less eager to re-establish normal relations with the Arab world. From the earliest days of Richard Nixon's presidency, the Nixon administration expressed deep apprehension over the unsolved Middle East conflict and its threat to world peace. The United States feared that a major clash between the superpowers might be precipitated by the Arab–Israeli conflict. Even before assuming office, President Nixon sent a personal emissary, William Scranton, to the Middle East to relay his intention of adopting an objective, 'even-handed' policy towards the disputing parties. In a meeting with the press on 27 January 1969 Nixon described the situation as explosive, 'a powder-keg' that, unless defused, might lead to a nuclear-powers confrontation, which he wished to avoid. Secretary of State Henry A. Kissenger said in a 23 June 1975 speech in Atlanta, Georgia, that in the Middle East 'an active American role is imperative':

(i)   because of our historical and moral commitment to the survival and well-being of Israel;

(ii) because of our important interests in the Arab world, an area of more than 150 million people sitting astride the world's largest oil reserves;

(iii) because the eruption of crisis in the Middle East would severely strain our relations with our allies in Europe and Japan;

(iv) because continuing instability risks a new international crisis over oil and a new setback to the world's hopes for economic recovery, threatening the well-being not only of the industrial world but of most nations of the globe;

(v)   because a crisis in the Middle East poses an inevitable risk of direct US–Soviet confrontation and has done so with increasing danger in every crisis since the beginning.[3]

Thus, the Ramadan war and the oil crisis provided a 'golden opportunity' for Washington to achieve Arab-American *rapprochement*. As one Middle East expert described it: 'The war provided the opportunity to [prove the disability of the Soviets] . . . while affording the occasion for a first demonstration of what the United States could do.'[4]

The Ramadan war set the stage for the United States to change its Middle East policy. Earlier, the joint Soviet-United States communiqué following Nixon's visits to Moscow in May 1972 had indicated the desire of both powers to relax the tensions that were building up in the Middle East as part of the general policy of *détente*. The Soviet Union's allies in the area were unhappy about this declaration, as they saw in it a weakening in Soviet support for their desire to rid their land of Israeli occupation. They also saw in it a *de facto* Soviet acceptance of the *status quo* as Israel continued its expansionist policies in the occupied territories.

Nixon's call for the recognition of Soviet influence and interests in Egypt following his return from Moscow was quickly interpreted in Arab capitals as a *quid pro quo* for Soviet acceptance of the consequences of Israel's aggression of 1967.

Détente survived the tensions created by the Ramadan war, and the United States and the Soviet Union went out of their way to emphasize this fact. Nixon regarded the outcome of the crisis as a 'victory for *détente*'. An editorial in the *New York Times* emphasized that during the October war, both the Soviet Union and the United States realized that they had a far more important stake in peace: 'Without *détente* we might have had a major confrontation in the Middle East, but with *détente* we have avoided it.' Joseph S. Szyliowicz described the American attitude during the October crisis as follows:

The US commitment to *détente* was so strong, however, that it proved remarkably understanding of Soviet actions. State Department officials pointed out that, while the USSR obviously approved of the Arab embargo – it did not mastermind the oil cutoff. Moreover, its Kremlinologists seemed convinced that Moscow was wary of the new power wielded by the Middle East states, over which the Soviet Union had little influence, as well as of the idea of increased Western co-operation in the face of the oil crisis. Finally, the State Department suggested that Moscow's dependence on Western technology for its internal development schemes made it unlikely that it would adopt policies discouraging trade and investment by the United States and other industrialized states.[5]

The United States' fundamental new approach was to re-establish American political and economic influence by encouraging 'moderation' among the Arab states. In Kissinger's view, stability in the area had to rest on the viability of moderate regimes with firm political bases. The Arab-Israeli impasse had to be solved on terms that gave those moderate regimes a stake, both economically and politically, and in terms that provided defense against more radical elements internally and elsewhere in the Arab world.

On 6 September 1973, for the first time, President Nixon publicly linked American oil needs to Middle Eastern policy. In a press conference he talked about the necessity of Congress acting on his domestic energy proposals. He argued:

> If the Congress does not act upon these proposals – which, in effect, has [*sic*] as their purpose increasing the domestic capacity of the United States to create its energy, it means that we will be at the mercy of the producers of oil in the Middle East . . . The problem that we have here is that as far as the Arab countries are concerned, the ones that are involved here, is that it's tied up with the Arab-Israeli dispute. That is why . . . we have put at the highest priority making some progress toward the settlement of the dispute.[7]

On 6 October 1973, only one month after the President's press conference and exactly two weeks after Kissinger had taken the oath of office as Secretary of State, the fourth round of the Arab-Israeli war erupted and the 'powder-keg' exploded. The oil war which followed proved effective in bringing about a change in the United States' peace-keeping efforts. At first, the United States declared that its foreign policy would not be changed because of pressure from the Arabs. Nevertheless, many American officials admitted, one way or another, the impact which the oil crisis had on American Middle East policy. This was even acknowledged by Robert W. Tucker, who advanced the possibility of direct American military intervention to occupy the Arab oil fields. Professor Tucker said:

> Although the sudden intrusion of the oil weapon cannot be said to have provided the initial promptings of the new policy, there is no question but that it gave this policy greatly added incentive and seemingly compelling logic. For the lesson widely drawn from the Arab embargo set off by the October war has been that a future war between Israel and the Arab states would in all probability provoke

another and more serious embargo.[8]

While the United States would not relinquish its commitment to Israel, it did not want to lose the initiative for its attempt to restore Arab confidence in the 'genuineness' of American efforts. At the same time, the structure of Soviet-American *détente* had somehow to be preserved. Thus, the Americans moved on two fronts — providing massive support to Israel to enable it to keep fighting, and orchestrating a military cease-fire which, by leaving neither side in a clearly suprerior position, created a situation in which it was possible to promote talks leading to disengagement on the Israeli-Egyptian fronts and, later, on the Israeli-Syrian fronts. The United States' so-called 'even-handed' policy was disclosed by Kissinger on 25 October when he said: 'The conditions that produced this war were clearly intolerable to the Arab nations and . . . in a process of negotiation it will be necessary to make substantial concessions.'[9]

The use of the oil weapon had many drastic consequences, not only on the Middle East, but on the world as a whole. Naturally, the United States was the main target of the oil embargo, given its unequivocal and unconditional support for Israel. D.A. Rostow, a leading White House adviser, explained that: 'The US has for long now been the only power that can pressure Israel. It is this power that precisely provoked the Arabs into using their oil weapon against us.'[10]

The Arabs maintained their hope that the United States would be persuaded to follow a balanced and even-handed policy, but they were to be frustrated: the Carter administration convinced Egypt's President Sadat to sign the Camp David peace treaty with Israel, a treaty which had been rejected by all the other Arab states for the simple reason that it failed to achieve the minimum demands of the Arabs and Palestinians. The Camp David Accords eroded the last ray of hope for achieving any breakthrough in Arab-American relations.

Five years have elapsed since the signing of the Camp David Accords, but there is still no hope of their leading to the achievement of lasting peace in the Middle East. The ability of such a treaty to establish a permanent peace even between Israel and Egypt is much in doubt. Egypt has been isolated by the rest of the Arab world because of its unilateral action as the only Arab country to sign a peace treaty with Israel. The Egyptian regime's inability to move forward politically to convince other parties to join in the Egyptian-Israeli 'peace-treaty' has led to an impasse. Indeed, many prominent American politicians and observers have asked the new American administration under President

Ronald Reagan to give up the Camp David framework and seek a new viable peace initiative that could be one of the major achievements recorded by the Reagan administration. Given the unanimous Arab rejection of the Camp David framework and Israel's intransigence and subsequent invasion of Lebanese territory and bombing of Iraq's nuclear reactor Osirak, there is no immediate hope for achieving any improvement in Arab-American relations. The Arabs remain convinced that their security is threatened by Israel and that Israel is not really interested in peace. In fact, continued aggressive acts by Israel, together with continued American support for Israel in terms of sophisticated weapons and financial aid, will undoubtedly widen the gap of misunderstanding between the United States and friendly countries in the area, especially the Kingdom of Saudi Arabia.

In a statement to the Saudi Press Agency on 8 August 1981 HRH Crown Prince (now King) Fahd bin Abdul Aziz, First Deputy Premier, stated:

> The US has swiftly and effectively co-operated with us and we appreciate that. However, . . . as for the general US policy towards the Middle East, we are not satisfied, especially in relation to the Palestinian issue and the rights of its people. I have made particular reference to the Palestinian issue and the rights of Palestinians, the root differences between the Kingdom and the US administration. This difference has been conveyed to all US administrations. This difference cannot be underestimated as it is directly linked to the security and stability of our region. It is also connected with American interests . . . We hope that the Reagan administration will admit the invalidity of the Camp David agreement as the framework for a just and overall peace in the Middle East and initiate a radical change in its Middle East policy which will lead to Israeli withdrawal from the Arab territories occupied in 1967 and establishment of an independent Palestinian state.

Prince Fahd established a set of guiding principles for the settlement of the conflict. They are:

(1) Israeli withdrawal from all Arab territories, including Jerusalem, occupied in 1967.

(2) The removal of Israeli settlements set up after 1967.

(3) The guarantee of freedom of worship in the holy places of Jerusalem.

(4) Recognition of the inalienable rights of the Palestinian people to return to their homeland or to compensation for those who do not wish to return.

(5) A provisional UN mandate for several months over the occupied West Bank and Gaza Strip.

(6) The establishment of the Palestinian state, with Jerusalem as its capital.

(7) The guarantee of the right of the Palestinian people and the people of the region to live in peace.

(8) The implementation of a peace plan to be overseen by the UN and some of its members.

Prince Fahd also pointed out that the implementation of these principles had the following pre-conditions:

(1) Ending America's unlimited support for Israel.

(2) Putting an end to Israeli arrogance, whose ugliest expression is represented by Israeli Premier Menachem Begin.

(3) Admitting that, as Yasser Arafat, the chairman of the PLO, has said, the Palestinian state is the basis of a peaceful Middle East settlement.

Israel's invasion of Lebanon in June 1982, the massacres at Shatila and Sabra in September 1982, and the Fahd Plan finally provoked the United States into making a major declaration in September 1982. Now known as the Reagan Plan, this declaration called for Israeli withdrawal from the West Bank and Gaza, the establishment of a confederation between Jordan and the evacuated occupied territories, and the recognition of Israel by the Arabs.

The Arab leaders, although appreciative of some positive aspects in the Reagan Plan (for example, the imposition of a freeze on new settlements and the call on Israel to withdraw from occupied territories), felt that the plan fell short of the minimum Arab demand − an independent Palestinian state. This position was codified in the Fez Summit Conference in late 1982. The Fez Delcaration made it clear that the Arabs are willing to live in peace with Israel, if Israel relinquishes the Arab lands it conquered militarily in 1967 and 1982. The Fez Declaration drew attention to the unacceptable asymmetries in the Reagan Plan. Under the plan, Palestinians were expected to surrender their rights to statehood by simply accepting a kind of loose confederation with Jordan, while Israel was asked to 'surrender' the West Bank and Gaza.

In effect, the Palestinians were expected to surrender their inherent right to sovereignty in their own historic land, while Israel was simply required to give up its bogus claims of 'sovereignty' in the Palestinian territories occupied in 1967.

One fact remains clear. The Arabs have been successful in defining a new framework that differs from the unacceptable Camp David one. The Reagan Plan could easily be interpreted as a response, albeit a weak one, to the Fahd Plan of 18 August 1981, and a move away from the Camp David framework.

A neutral evaluation of the peace process in the Middle East can be found in President Nicolae Ceausescu of Romania's view of the conflict. In an interview published in *Newsweek* on 16 May 1983, Ceauşescu said:

As is known, there are several proposals: the Fez proposals of the Arab countries, the proposals of President Reagan, the proposals of Romania for an international conference.

I think that the necessity to organize negotiations in which the Palestine Liberation Organization should be an active participant must be taken as a starting point. Without the participation of the PLO, no negotiations can be organized . . . Negotiations concerning the Palestinians can be conducted with nobody but them . . .

Opportunities have been missed before, and if this one is missed too, it will be hard to say that things will remain the same for a long time.

Increased American inerest in the area's problems and the eclipse of Soviet influence have resulted in a number of changes in American Middle East policy. These changing political parameters can be examined under three headings: safeguarding oil supplies, the preservation of the general security of the area, and the resolution of the Arab-Israeli conflict.

Safeguarding Oil Supplies

It is now obvious that the United States' Middle East policy is influenced by the American and world energy situation. With foreign oil constituting 41 percent of domestic availability and Arab oil accounting for about 47.3 percent of total imports (see Tables 15.1 and 15.2 and Figure 15.1), the United States can no longer ignore its increasing

Table 15.1: Crude Oil Balances in Selected OECD Countries, 1976 and 1980 (thousand metric tons)

	United States	France	Federal Republic of Germany	Japan	OECD Europe	OECD Total
1976a						
Production	455,522	1,057	5,524	576	40,878	592,946
Imports	308,140	121,143	105,291	229,643	647,496	1,233,064
Inventories	−2,460	−576	−2,536	−376	−3,646	−646
Domestic availability	726,866	122,624	108,248	249,107	665,164	1,757,591
Imports/production	0.68	114.6	19.1	398.7	15.8	2.1
Imports/Domestic availability	0.42	0.99	0.97	0.92	0.97	0.70
1980						
Production	426,354	1,491	4,613	430	117,382	639,445
Imports	285,252	109,495	97,920	215,567	559,824	1,100,026
Inventories	−3,649	−279	−2,911	−4,076	−6,922	−15,491
Domestic availability	703,160	110,707	99,552	211,921	608,084	1,646,262
Imports/Production	0.67	73.4	21.2	501.3	4.8	1.7
Imports/Domestic availability	0.41	0.99	0.98	1.02	0.92	0.67

Notes: a. Data for 1976 include crude oil, natural gas and feedstock, whereas data for 1980 include crude oil.
Source: OECD, International Energy Agency, *Energy Statistics, 1976-1980* (OECD, Paris, 1982).

Table 15.2: Crude Oil Trade, 1980 (thousand barrels per day)

From/To	United States	Japan	France	West Germany	Total Europe
Bahrain	—	—	—	—	—
Iraq	31.4	370.4	483.7	51.7	1,230.6
Kuwait	34.2	166.5	63.3	16.3	391.4
Oman	26.6	145.6	1.7	25.3	49.8
Qatar	19.4	134.4	53.0	4.2	198.7
Saudi Arabia	1,241.4	1,573.9	779.8	492.1	3,734.3
Syria	1.7	—	4.3	7.4	87.3
UAE	180.6	646.7	151.9	123.4	615.2
Algeria	447.4	29.4	95.8	134.4	615.2
Egypt	33.9	3.2	1.0	4.6	257.4
Libya	539.1	23.8	40.2	308.6	782.7
Tunisia	4.8	—	3.0	8.1	76.0
Total Arab	2,555.7	3,093.9	1,677.7	1,176.6	7,764.7
Total World	5,402.3	4,595.0	2,219.3	1,963.1	10,977.0
Percentage Arab/World	47.3	67.3	75.6	59.9	70.7

Source: Energy Economics Research Ltd., *World Oil Trade* (Dec. 1981), pp. 6 and 7.

dependence on the Arab world for the bulk of its oil requirements. Equally important for the United States is the even more serious energy outlook in Japan and Europe. Japan depends on foreign imports of oil for 67 percent of its domestic availability, and on those sources for about two-thirds of its total imports (see Tables 15.1 and 15.2). For Europe as a whole, the situation is perhaps more drastic, with oil imports constituting 92 percent of domestic European availability and a dependence on Arab sources for over 70 percent of total imports.

The West's reaction to the possible interruption of oil supplies differ from one country to the next. Yet all Western countries share a common concern to mitigate the consequences of possible interruptions in supplies. President Carter underscored the United States' position by declaring that world dependence on Arab oil is expected to rise, and that any interruption of supplies from the region would constitute a threat to national security and would likely precipitate an economic crisis of graver implications than the Great Depression of the 1930s. He pointed out two ominous dangers that might call for an immediate American military response to protect US 'vital interests' in the region — political instability in the region, and Soviet encroachment on it.

Western reaction in general has taken two principal courses. The first emphasizes that security of oil supplies is contingent upon Western

Figure 15.1: American Oil Imports (million barrels per day)

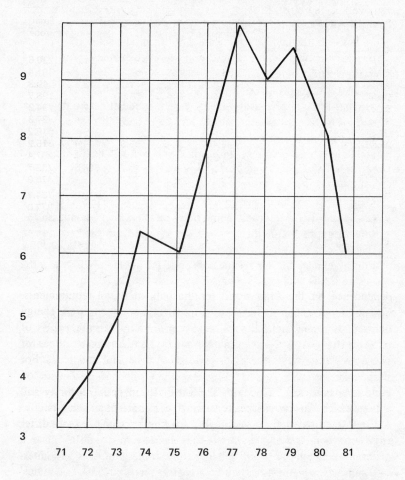

energy policies designed to reduce oil imports from the Middle East through conservation and minimization of waste. During this transitional period, efforts should be directed at developing alternative energy supplies, while at the same time maintaining negotiations with oil producers in the common interests of both parties. This position is predicated on the following observations about the nature of oil and its potential displacement by other energy sources:

(1) Since Middle East oil is necessary for the world economy and will remain so for some time to come, oil consumers must ensure its

steady flow to meet their basic needs.

(2) Since the current dependence on Middle East oil is potentially threatening because decisions about oil supplies or oil prices are no longer in the hands of the United States and its allies, it is imperative for the West to stabilize the increasing dependence on this oil in the shortest time possible.

(3) Since the only possible option to secure the necessary energy supplies is a comprehensive energy policy aimed at developing alternatives to oil, and alternatives to Middle East oil in particular, and the optimum use of energy sources to extend their life-span to the maximum possible extent, the United States should act immediately to formulate a joint energy strategy with Europe and Japan.

(4) The demand for Middle East oil should be reduced, so as to augment the oil-consuming countries' bargaining power *vis-à-vis* OPEC.

(5) Since oil is a tradeable commodity and its flow depends on the flow of goods, technology and investment opportunities in and from the oil-consuming countries, co-operation could replace confrontation by expanding the avenues of economic interaction between the oil producers and oil consumers.

The second basic position has emphasized military options. Proponents of this view argue that military confrontation is the most effective way to neutralize the threat of Arab oil embargoes. At the same time, the West must freeze its exports of food and spare parts to the Middle East.

In an article on 7 October 1974 entitled 'Thinking the Unthinkable', the editor of *Newsweek* concluded that the three most talked about options then were: (i) psychological warfare to unnerve the enemy; (ii) covert operations aimed at terrorizing Arab sheikhs; and (iii) military intervention in the form of American parachute assaults – supported by similar Israeli moves. A study prepared for the House Committee on International Relations entitled *Oil Fields as Military Objectives: A Feasibility Study* is more evidence of this position.[11] To this must be added the intervention scenarios of Tucker and Ignotus and many others.

The feasibility and desirability of this position are questionable on several grounds:

World public opinion is no longer willing to accept colonialist activities and imperialist designs. The Third World is particularly conscious of its collective security, its quest for independence, and its inalienable right to self-determination. The Europeans, with a long history of colonialism, are particularly sensitive about their tragic past relationship with the Third World. Finally, the American public is still reeling from the post-Vietnam syndrome and is not likely to accept a new front.

The Third World is no longer docile, as it was in colonial days. Any military adventure by the West is likely to meet stiff and crippling resistance. The Vietnam experience is still a vivid reminder of this scenario.

The precarious balance of terror that mains world peace could not be expected to be maintained in the face of a military campaign against OPEC or OAPEC. Western military activity in the Middle East would most likely provoke a military confrontation between the superpowers. The annals of the Ramdan war reveal clearly the severe strains that developed between the United States and the Soviet Union, and the risk of confrontation between the two powers would be much greater if the United States were to attempt to seize control of the oilfields in the Middle East.

The Arab oil producers are cognizant of the importance the West attaches to its oil supplies. They are equally desirous of smooth and orderly interaction with Western economies. They need Western technology, goods and markets. But the relationships must be of mutual interest, and between equals. Gone are the days in which the West could simply dictate terms. *Pravda* was very clear on this issue when A. Vasiliev was quoted as saying:

> The military option is ineffective in securing oil supplies from the Middle and Near East and from the Arabian Gulf Countries. But peace, refraining from interference in the domestic affairs of the area, and balanced and just relationships with the oil producers can precisely be the conditions that would ensure the steady flow of oil to the capitalist and developing countries. The resolution of the oil crisis, or for that matter the 'resources crisis', requires peace and voluntary economic co-operation beneficial to both sides and not rash military adventures.[12]

Maintenance of the Area's Security

For the United States, the question of security in the Middle East is directly correlated with the presence and influence of the Soviet Union in the area. It is, therefore, a top priority for American foreign policy to reduce and contain the Soviet presence. Although America takes for granted the influence of the Soviet Union in the area and the impossibility of its exclusion, United States foreign policy endeavors to contain it within boundaries that do not threaten vital Western interests in the area such as oil.

Soviet strategy in the Middle East, on the other hand, operates on four axes: first, the Islamic axis, which encompasses Islamic but non-Arab countries that stretch along the southern border of the Soviet Union (Turkey, Iran and Afghanistan); secondly, the Arab-Israeli axis, which includes the confrontation states (Syria, Lebanon, Jordan and previously Egypt); thirdly, the Arabian Gulf axis, which encompasses the oil-producing states along the Arabian Gulf; and fourthly, the North African axis, with Libya the focus of Soviet attention.

Soviet influence has risen in direct proportion to Western mistakes, particularly those committed in the 1950s, beginning with the Egyptian arms deal débâcle of 1955. The Soviet presence was firmly established during the 1967 June war, because of the USSR's monopoly position as the Arabs' only arms supplier. Soviet-Arab relations cooled in 1973 in the wake of the Ramadan war, when: '. . . Sadat turned towards Washington. During the Spring of 1976, the last Soviet military base in Egypt was dismantled . . . and in less than five years a massive presence was totally liquidated.'

Recent Western mistakes in the Arab world and in Africa have given the Soviets an opportunity to find a replacement for Egypt in the area. The Soviet presence in nearby countries raises the likely eventuality of its return to Egypt and to other areas from which it was dislodged. This possibility becomes more probable, the more mistakes the West makes in the area. Unfortunately, these have been and continue to be substantial.

The United States however, has a golden opportunity to lay to rest the possibility of Soviet resurgence in the Middle East. This calls for developing and up-grading Arab-American relationships. What prevents this from happening is the Arab-Israeli conflict.

The Resolution of Arab-Israeli Conflict

The progress and development of the entire Middle East is contingent upon resolving the central conflict between the Arabs and Israel. The United States must play a major role in finding a just and acceptable solution to the Palestinian question.

Regardless of what Israel represents to the United States and the 'importance' of the assumed 'gains' that America mistakenly believes it derives from its special relationship with Israel, the Palestinian question is going to be the major determinant of American interests in the Arab world. Without a just and lasting solution to this issue, American interests will remain vulnerable and precarious, and Arab-American relations will never be stable or lasting.

If the past is any indication of what the future will bring, it is abundantly clear that with the continuation and escalation of the conflict between the Arabs and Israel, the influence of the United States in the Middle East will most likely be eclipsed and the Soviet Union will eventually capitalize on American faults. The United States' support for Israeli aggression (or its inability to prevent this aggression, as the events in Lebanon in 1982 clearly reveal) is likely to dash the hopes that the Arabs continue to hold for improving Arab-American relations.

This position has been echoed in the Report of the Special Task Force on the Middle East, constituted by the Atlantic Council of the United States from 30 American experts on military, diplomatic and security affairs. The Report concludes:

Present discussions and negotiations regarding the future of the Palestinian question, and other aspects of the Arab-Israeli conflict, will perhaps subject Arab-Israeli relations to severe strains. American support for Israel must not be automatic, because Israeli aggressive policies – the same goes for those of the Arabs – may be in conflict with US interests in the area. We believe that the US should not leave any shadow of doubt about its commitment to Israel's independence and security and about its unwillingness to allow this commitment – aside from the necessary security arrangements that are part of the settlement itself – to be anything but a parameter of the American position. We also believe that the US must declare with equal clarity its adherence to the UN resolution 242 and its conviction that Israel's security can be served better through peace with the Arabs, including the Palestinians in Gaza and the West Bank, than through the continuation of war and terrorism. Although it is also necessary to balance US commitment for Israel's security against its vital interest in establishing good relations with several countries in the Arab world, it is important that neither objective should be sacrificed for the other.

Despite the Reagan administration's repeated declarations of its opposition to the PLO and a Palestinian state, there are a number of indications that a new American position is evolving on this issue. First, there were the leaked press reports of Senator Percy's (Chairman of the Senate Foreign Relations Committee) statements that the United States supports the establishment of a Palestinian state under the leadership of Chairman Arafat. Then, there was the call by Zbigniew Brzezinski (former National Security Adviser) for American recognition of the PLO and for an immediate dialogue between the two parties. These moves culminated in the Reagan Plan of September 1982, which, as mentioned previously, called for an Israeli withdrawal from the West Bank and Gaza, the establishment of a confederation between Jordan and the evacuated Palestinian territory, and recognition of Israel by the Arabs.

American policy appears to be striving for some balance in its dealings with the Middle East. Thus far, the balance remains heavily tilted in Israel's favor. The Arabs expect more from the Americans. The new American position revolves around the following principles:

(1) Israel is a 'strategic base' for United States global security, and the United States is committed to Israel's existence and security.

(2) It is necessary to improve United States relations with the Arab oil-producing countries, and the protection and security of oil-wells and oil routes is best served through solid United States–Saudi relations.

(3) As a compromise between the two principles mentioned above, the United States is striving to define the extent of its support for Israel, by restricting its commitment to Israel's existence but not to Israel's expansion.

(4) In order to achieve the third objective, the United States must work towards a settlement of the Palestinian question that is acceptable to the PLO and the Arabs; this means side-stepping Egypt and the Camp David Accords and working with the other Arab countries as a joint body.

We turn next to the question of the nature of Arab oil power. This power, in our opinion, is the basis of much of what has happened in the way of recent Western accommodation or understanding of Arab positions.

Notes

1. Ghazi al-Gosaibi, 'American–Saudi Relations', a presentation to the Council on Foreign Relations, New York, 7 April 1976.

2. Mohamed H. Heikal, *The Road to Ramadan* (Quadrangle Books, New York, 1975), p. 115.

3. Henry Kissinger, speech delivered in Atlanta, Georgia, 23 June 1975.

4. Robert W. Tucker, 'Israel and the United States: From Dependence to Nuclear Weapons', *Commentary* (Nov. 1975), p. 29.

5. Joseph E. Szyliowicz and Bard E. O'Neill (eds.), *The Energy Crisis and US Foreign Policy* (Praeger, New York, 1975), p. 197.

6. Joe Stork, *Middle East Oil and the Energy Crisis* (Monthly Review Press, New York, 1975), pp. 236–7.

7. Quoted in Ruth S. Knowles, *America's Oil Famine* (University of Oklahoma, Norman, Oklahoma, 1980).

8. Tucker, 'Israel and the United States', p. 29.

9. *New York Times*, 26 Oct. 1973.

10. John Collins and Clyde Mark, *Oil Fields as Military Objectives: A Feasibility Study* (Washington, D.C., 1975).

11. Ibid.

12. A. Vasiliev, 'Oil and False Imperial Dreams: A Soviet Viewpoint', translated into Arabic from the Soviet daily, *Pravda*.

PART FIVE
ARAB PETROPOWER: AN EVALUATION

16 THE NATURE AND SUBSTANCE OF ARAB OIL POWER

All political attitudes have points of weakness and strength. The weaknesses are the source of defeat. The strengths indicate the ways and means which, during a confrontation, permit one party to inflict defeat on the other. Governments naturally attempt to eliminate all points of weakness, or to reduce their impact. On the other hand, power is cherished, developed and enhanced.

The political status of oil — petropolitics[1] — is no different. It suffers from points of weakness and enjoys elements of strength. Professor Hamed Rabie believes that a certain key factor, by itself, may inspire victory or inflict defeat. Furthermore, sources of weakness and strength vary over time and in their importance as decisive factors. The first rule of political activity centers on the methods of dealing with these factors, not only for the first party but also for adversaries in a state of confrontation.

Professor Rabie summarizes the rules of political success in five principles:[2]

(1) the ability to distinguish, in any given political position, between points of weakness and points of strength;
(2) the maximum exploitation of points of strength;
(3) the neutralization of points of weakness, and their conversion into points of strength whenever possible;
(4) the maximum exploitation of the enemy's points of weakness;
(5) the attempt to turn the enemy's points of strength into points of weakness.

To maximize its political influence as an important factor in international politics, Arab oil should take these five principles into consideration. The purpose of this section of the book is to explore both the potential of oil as a political weapon and the future prospects of the Arab-Israeli conflict. In doing so, the following questions will be addressed:[3]

(1) Have shifts in the balance of power occurred between the industrialized nations and the oil producers? If so, how real is the power

167

that the oil producers have assumed?

(2) What is the extent of the shift in power, and how far could the Arab exporting countries go in using their newly acquired power?

(3) Where does oil stand in relation to other forms of energy, and what is the future of oil as a vital source of energy?

(4) Do the Arab states have the willingness and ability to reward accommodating states, and punish unco-operative ones?

(5) Do the accommodating states gain more from adjusting their policies to satisfy the Arab demands than by not doing so?

(6) Can the Arabs afford the economic and political costs of administering 'punishment' to non-accommodating states? Are such costs likely to be compensated for by the gains from any political accommodation which the attempt to 'punish' may produce?

(7) Can an exchange of political for economic advantages between the Arabs and other states really be achieved and maintained? Does a mechanism exist to ensure that, once agreement is reached, other states will keep their side of the agreement with the Arabs?

(8) Which economic tool or group of tools is most efficient in bringing about political accommodation? Does there exist an economic instrument, or instruments, that the Arabs can use to derive political dividends with impunity?

The answers to the above questions will determine whether the Arabs can wield their 'tactical economic instruments' to gain political accommodation and acceptability for their causes, and the extent to which the Arabs can maintain a position of influence in the international arena. As a general principle, the extent of political accommodation by other states will be greater, the more demonstrable the willingness and ability of the Arabs to reward accommodation, the higher the economic cost of 'punishment' for non-accommodating states, the lower the domestic cost of punitive policies for the Arabs, and the more efficient the mechanisms to ensure sustained compliance with Arab goals on the part of other states.

The Quest for Power

Power plays a major role in international relations. The quest for power has always been an important factor in motivating human activities and in determining one man's relations with another. So, too, power has become the distinguishing feature in formulating relations among different nations.

Man's desire for life, supremacy and dominance, as natural instincts, have been common denominators for all. The struggle for power is therefore a universal principle that cannot be defied or ignored. Professor Morgenthau, one of the leading authorities on international relations and advocate of the realism school of politics, defines power as 'Man's control of the minds and actions of others'. In brief, Professor Morgenthau regards power as the essential factor in international relations, if not the only one:

International politics, like all politics, is a struggle for power. Whatever the ultimate aim of international politics, power is always the immediate aim. Statesmen and people may ultimately seek freedom, security, prosperity, or power itself. They may define their goal in terms of a religious, philosophic, economic, or social ideal. They may hope that this ideal will materialize through its own inner force, through divine intervention, or through the natural development of human affairs. They may also try to further its realization through non-political means, such as technical co-operation with other nations or international organizations. But whenever they strive to realize their goal by means of international politics, they do so by striving for power.[4]

The influence of power has left an important mark on both the past and the present. Whatever the declared final goals of policy-makers, the accumulation of power and eventual dominance seem to be the ultimate aims. The principle of power is undoubtedly a key element in the study of international politics.

Nevertheless, lines have to be drawn between political power and the exercise of force. The physical side of political power can be translated into reality through war, for example, in which case political means are abandoned in favor of military or paramilitary force. Our subject, however, is the political use of power – political power – and not its military face.

Political Power

Political power is a psychological relationship between those exercising it and those subjected to its influence. It provides those exercising power with a degree of control over some of the actions of those subjected to its influence through three sources: expectation of reward, fear of harm, and love or respect of the discipline. The exercise of power is carried out by order, threat or conviction, or by a mixture of two or more of these factors.

In the international sphere, the success of the relationship between the country exercising influence and the country subjected to it is attained only when such influence produces positive results for the influencing country. Otherwise, if unsuccessful, such influence would not represent political power at all. The United States, for example, exercises political power over Puerto Rico, because the behavior of the government and people of the island are consistent with the wishes of the United States government. In general, we say that 'A' is exercising political influence over 'B' when 'A' has the ability to control some of the actions of 'B' by influencing his mind or way of thinking.

Arab Oil Power

The source of Arab oil power depends on the political power of the Arab nation, which is in turn dependent on the economic position of Arab oil-producers. Oil is essentially a commercial commodity, and oil exchange is basically a commercial operation. But the use of the 'oil weapon' implies that both the commodity and the commercial exchange are employed for purely political purposes. The economic power inherent in a position of monopoly, whether commanded by seller or buyer, creates a great deal of political power. It follows that there is a fundamental difference between economic policies followed for 'natural' commercial purposes, and those followed as specific tools to achieve certain political goals. This distinction becomes paramount when discussing economic policies in international affairs, when certain economic policies may be used as a means to achieve the wider aim of controlling the policies of other countries.

Economic or financial policies are often subject to adjustment or correction by virtue of specifically economic or financial criteria. For example, what will be the effects on savings or expenditure of raising the salaries of public servants? What would be the result on the general levels of prices of levying higher tariffs on imported goods? Will particular policies be beneficial economically or financially? The decisions taken to correct or re-evaluate such economic policies are generally considered within specific economies or financial criteria. But when a certain policy is adopted expressly for the purpose of increasing the influence of one country over another, it becomes essential to judge these policies and their purposes by a different set of criteria, beginning with their contribution to the national power of the policy-making country. Such policies may be pursued because of a specific political goal even if they are unjustifiable from a purely economic point of view.

The essence of Arab oil power is based on the simple economic truth of the law of supply and demand, and is ultimately dependent on the commercial relations between the oil producers and consumers. Since the early 1970s, a change has occurred in the balance of trading power between the groups. Power in the international oil markets has shifted from the multinational oil companies and the major oil importers to the producers.

The principal oil-importing countries of Western Europe and Japan are now exposed to the uncertainties of future oil policies. In 1974 Western Europe imported an estimated 85 percent of its total oil requirements from the oil producers in the Middle East, and Japan imported approximately 90 percent. At the beginning of 1974 United States imports were but 18 percent, but by the end of the year imports had jumped to 28 percent of total oil requirements.

Even in the United States, the ratio of imports to production was as high as 67 percent in 1980, a year when imports were exceptionally low. As a percentage of United States total domestic availability in 1980, imports of crude oil represented 41 percent, with Arab oil accounting for over 47 percent of the total oil import requirements (see Table 16.1).

The European and Japanese situations are even more representative of the heavy dependence of OECD countries on oil imports in general (see Table 15.1) and on Arab oil in particular (see Table 15.2). Imports of oil accounted for over 92 percent of total domestic oil availability in European OECD countries in 1980: the ratio was 90 percent for France, 98 percent for West Germany, and over 102 percent for Japan. More importantly, 70.7 percent of the total European imports of oil came from the Arab world in 1980. In France this ratio was over 75 percent.

As of 1 January 1978 the Arab countries had 333.2 billion barrels of reserves, or about 52 percent of the world total. In 1977 the Arab share of total world crude oil production stood at about 32 percent, and accounted for over 62 percent of total OPEC daily production. By 1980 the Arabs' share of the total world production of crude had increased to 36.3 percent and their share of OPEC's production to 71.7 percent (see Table 16.1). Their share of total world exports of

Table 16.1: Oil Production and Reserves

	1970	1975	Production 1977 (thousand barrels per day)	1980a	1981	Reserves 1 January 1978 (million barrels)	Ratio of reserves 1978 to annual production 1977
Arab countries	14,146.3	16,343.0	18,979.0	19,503.0	16,065.0	333,245	48.11
Algeria	1,008.4	946.0	990.0	1,016.0	812.0	6,600	18.26
Bahrain	76.6	61.0	54.0	48.0	44.0	270	13.71
Iraq	1,566.2	2,240.0	2,150.0	2,638.0	917.0	34,500	43.96
Kuwait	2,734.5	1,807.0	1,700.0	1,382.0	939.0	67,000	107.98
Libya	3,321.6	1,488.0	2,050.0	1,785.0	1,104.0	25,000	33.41
Oman	332.4	342.0	350.0	283.0	319.0	5,650	44.21
Qatar	363.3	441.0	350.0	472.0	405.0	5,600	43.82
Saudi Arabia	3,548.9	6,827.0	8,950.0	9,630.0	9,641.0	150,000	45.92
UAE	693.8	1,695.0	2,030.0	1,709.0	1,513.0	32,425	43.76
Neutral Zone	500.6	496.0	355.0	540.0	371.0	6,200	47.84
Other OPEC	9,084.9	11,155.0	12,045.0	7,669.0	6,839.0	112,590	25.61
Non-OPEC	13,874.0	13,836.5	15,223.8	11,973.0	18,516.0	102,162	18.39
Communist countries	7,916.5	11,822.2	13,123.7	14,569.0	14,596.0	98,000	20.46
World total	45,021.7	53,156.7	59,371.5	53,714.0	56,016.0	645,997	29.81

Notes: a. Twelve-month averages.
Source: International Petroleum Encyclopedia, 1978 and Oil and Gas Journal, 8 March 1982, p. 331.

Table 16.2: Arab Countries, United States, Japan, EEC and World Trade Matrix, 1980 (million US dollars)

Imports from/ Exports to	EEC	US	Japan	Arabs	World
EEC	—	38,329	6,382	53,238	670,118
US	54,601	—	20,790	13,415	220,703
Japan	18,286	32,961	—	13,138	130,460
Arabs	77,168	31,906	41,744	—	227,058
World	733,357	252,995	141,284	122,686	—
			(Row percentages)		
EEC	—	5.72	0.95	8.00	100.00
US	24.74	—	9.42	6.01	100.00
Japan	14.02	25.27	—	10.02	100.00
Arabs	33.40	14.04	18.38	—	100.00
World	58.64	20.23	11.30	9.81	100.00
			(Column percentages)		
EEC	—	15.15	4.52	43.39	53.37
US	7.44	—	14.71	10.82	17.70
Japan	2.49	13.03	—	10.65	10.45
Arabs	10.52	12.60	29.54	—	18.18
World	100.00	100.00	100.00	100.00	100.00

Source: *United Nations, International Monetary Fund, Directory of Trade Statistic Yearbook, 1981.*

oil ranged between 83 and 87 percent during the 1970s.[5]

Given the present relatively low price elasticity of demand for oil, the real question facing Arab OPEC producers is why they do not lower production today, charge higher prices, and have both more money today and more oil in the ground tomorrow. OPEC producers' mistrust of Western capital markets, following the Iranian and Argentinian episodes, should even cause them to add risk premiums to the prices they charge, reducing production still further and 'saving' it for the future.[6]

However, Arab OPEC producers appear unwilling to slow the flow of oil as this may jeopardize Western economies. Some Western observers argue that this is due to Arab dependence on Western markets for food, technology and investment,[7] but this is only a partial picture of the issues involved. The net economic advantages of slowing oil production are not necessarily unfavorable to the Arabs. Although Arab oil is still the main source of energy for OECD countries, OECD countries need not be the main source of food and technology for the Arabs. Domestic production and/or procurement from alternative sources is an Arab option, but alternative sources of energy are not readily available to most OECD countries. The ratios of oil imports to domestic oil production in the OECD countries displayed in Table 16.2 reveal these countries' significant dependence on imports, which, given the present allocation of proven reserves, must be procured from Arab sources.

Furthermore, the dependence of OECD countries on imported oil, and the increased role of Arab oil in meeting OECD oil requirements, are not fully reflected at the political level. In other words, either the Arabs have not fully exploited their oil leverage, economically and politically, or the Western countries have not yet compensated the Arabs for their excess supplies of oil at less than maximum profit prices.

Although Arab oil revenues could theoretically be higher than they are now, they are formidable nonetheless. In 1980 Arab oil revenues reached a peak of about $208.3 billion; ten years earlier they were a meagre $4.6 billion (see Table 4.1). This 50-fold increase in revenue has catapulted the Arabs into the role of a major center of purchasing power at a time when most of the world's economies are either stagnant or contracting.

These substantial revenues are being used increasingly to finance the massive development effort in the Arab world. However, given the limited industrial absorptive capacity of the Arab domestic economies, these revenues are still mainly used to cover mounting import bills, investment abroad, and aid to the developmental effort of the Third World.

The Trade Stakes

In 1980 Arab world exports exceeded $227.1 billion, or over 18 percent of total world exports. During the same year, Arab imports were $122.7 billion, or over 9.8 percent of total world imports.

These figures compare rather favorably with the corresponding percentages for countries as large as the United States and Jordan. In fact, total Arab exports exceeded American exports in 1980. Arab imports, however, were only about half total United States imports in that year. Arab world exports were almost twice as large as, and Arab imports almost equal to, the corresponding Japanese figures (see Table 16.2).

Japan obtained over 28.8 percent of its total imports from the Arab world, and delivered over 10 percent of its total exports to that area in 1981. The EEC countries delivered over 8 percent of their total exports, and obtained over 10.5 percent of their total imports from the Arabs in 1980. Even the United States delivered over 6 percent of its total exports to the Arab area and obtained over 12.6 percent of its total imports from there in 1980. The other side of this picture reveals the Arabs importing over 43 percent of their total import bill from the EEC group and about 11 percent from each of the United States and Japan in 1980.

The picture is clear: over 65 percent of total Arab exports (mainly oil) are destined to the United States, Japan and the EEC group, and about 65 percent of Arab imports are purchased from there. This enormous volume of trade (excluding the lucrative invisible trade) theoretically should translate itself into wider co-operation, as indeed is the case among major trading partners – for example, the co-operation between the EEC, the United States and Japan. However, this has not yet taken place as far as the Arab countries are concerned, despite the fact that much of Arab trade is handled directly or indirectly by Arab governments.

It is equally important to examine the increased Arab trade with oil-importing developing countries (OIDCs). Arab OPEC members imported almost $12.4 billion worth of goods from this group in 1980. The corresponding figures for 1978 are almost half this figure ($6.8 billion), and as early as 1974 the figure was less than $3.4 billion. Total Arab imports from these countries exceeded $17.2 billion in 1980, which represents a more than three-fold increase since 1974 (see Table 16.3). Saudi Arabia's imports from OIDCs increased from $586 million in 1973 to $4,384 million in 1980. To this we must add the impressive gains in the consultancy, construction and engineering contracts won by OIDCs in the Arab world in recent years. For instance, over $9 billion worth of contracts were awarded to South Korean firms by the Arab world in 1978 alone.[8] Similar gains have recently been registered by Brazil, India, Taiwan and Turkey. Although these increases in Arab imports from OIDCs are impressive, the absolute values are nonetheless low, less than 14 percent of total Arab imports in 1980.

Remittances and Development Aid

The number of migrant workers in the West Asian Arab oil-producing states is estimated to have exceeded 2.3 million in 1981. This figure represented 10 percent of the total labor force and around 3 percent of the total population in the region. Excluding Iraq, the number of

Table 16.3: Arab Imports from Oil-Importing Developing Countries
(million US dollars)

	1974	1978	1980
Algeria	419	898	737
Bahrain	96	218	233
Egypt	436	1,162	1,030
Iraq	648	534	1,706
Jordan	131.5	350.4	405.5
Kuwait	325	894	1,338
Lebanon	510.8	486	859
Libya	569	708	1,277
Morocco	269	273	356
Oman	59	178	293
Qatar	59	126	170
Saudi Arabia	948	2,569	4,384
Sudan	145	177	326
Syria	310	568	787
Tunisia	149	221	281
UAE	375	934	2,448
YAR	71	334	401
Yemen PDR	64	136	197
Total	5,585	10,786.4	17,228.5
Arab OPEC total	3,702	6,861	12,353

Source: United Nations, International Monetary Fund, *Directory of Trade Statistics Yearbook, 1981.*

migrant workers represented over 32 percent of the total national population of the oil-producing countries of the region.[9]

The remittances these workers send home were estimated by the *International Monetary Fund* (IMF) to have reached $10 billion in 1980, whereas they were $1.3 billion in 1974 and $5.4 billion in 1977.[10] These flows covered a major proportion of the visible and invisible import bills of several countries, including Pakistan, Bangladesh and India.

In addition to these remittances, the Arab members of OPEC have been extremely generous with their aid to developing countries in a variety of ways — official direct aid, concessional aid, and other types including private donations. In 1980 official Arab direct aid is estimated to have reached $6,798 (see Table 2.5), whereas concessional aid has been placed at $6,802 million.[11] This adds up to a staggering $14 billion aid portfolio.

Arab official and concessional aid to developing countries compares vary favorably with OECD aid. Saudi Arabia's official aid was almost as high as that of Japan or West Germany in 1980. When aid from Saudi

Arabia, Kuwait and UAE aid is combined, it ranks second only to the United States. If aid is measured as a percentage of GNP, then Arab aid is far more significant than OECD aid. As a case in point, Kuwait's official aid was as high as 10 percent of its GNP in 1977. In the same year, the average percentage for the Arab members of OPEC was 2.34 percent in 1980 as against 0.27 percent for the United States.[12]

The Arab Petrodollar Bonanza

The high volume and value of oil exports, at a time when the import absorptive capacity of Arab economies is limited, necessarily implies decisions to invest abroad. These accumulated surpluses, at best, represent no more than a future claim on imports and/or foreign assets. They need not, however, be denominated in any particular currency and they need not be accumulated in any one particular country or in a specific asset. However, three characteristics dominate Arab petrodollar surpluses:

1. Four Arab countries – Saudi Arabia, the UAE, Libya and Qatar – account for the major part of the Arab petrodollar surplus and, for that matter, of the total OPEC surplus.
2. The Arab petrodollar surplus is primarily held in liquid wealth instruments – in 1980, for instance, bank deposits, treasury bills or bonds and holdings at the IMF and the International Bank for Reconstruction and Development (IBRD) added up to $61.7 billion – and the rest is held in quasi-liquid portfolio investments, so that the surplus (as shown in Table 16.4) is in effect a total net cash surplus.
3. The lion's share of Arab petrodollar surpluses are held directly, or indirectly through foreign branches of domestic banks, in the United States and the United Kingdom; estimates of the total value of Arab holdings in the United States range from $200 to $400 billion.[13] The strict government disclosure requirements in the United States and the 1979 freezing of Iranian assets appear, however, to have had only a marginal impact on the placement of Arab surpluses in America. According to the Bank for International Settlements in Basel, Switzerland, the 13 nations of OPEC are continuing to increase their proportion of non-dollar assets, but only gradually. Moreover, OPEC holdings of United States Treasury bills, notes and bonds increased by $17.8 billion in the 21-month period beginning 1 January 1980.[14]

Table 16.4: Identified Deployment of OPEC Surpluses (million US dollars)

	1980	1st half 1981
United Kingdom		
Sterling bank deposits	1.4	0.6
Eurocurrency bank deposits	14.8	4.9
British government stocks	1.9	0.4
Treasury bills	−0.1	0.2
Other sterling placements	0.1	—
Other foreign currency placements	−0.5	−0.5
	17.6	5.6
United States		
Bank deposits	−1.1	−0.5
Treasury bonds and notes	8.2	5.5
Treasury bills	1.4	0.2
Other portfolio investment	4.7	2.7
Other	0.9	−0.4
	14.1	7.5
Bank deposits in other industrialized countries	26.2	1.8
Other investments in other industrialized countries[a]	17.0	11.1
IMF and IBRD[b]	4.9	1.3
Loans to developing countries	6.7	2.9
Total identified deployed net cash surplus	86.5	30.2
Residual of unidentified items[c]	32.5	6.8
Total net cash surplus derived from current account[d]	119.0	37.0

Notes: a. Mainly loans and holdings of equities.
b. Includes holdings of gold.
c. The residual may reflect errors in either the current or capital account.
d. Includes net external borrowing of $9 billion.
Source: Bank of England, *Quarterly Bulletin*, vol. 22, no. 1 (March 1982), p. 35.

The concentration of Arab petrodollars in a few countries, and their allocation to even fewer financial centers and over a limited range of instruments, have several implications. First, most of the surplus funds (denominated in American dollars) have been transferred from European control to Arab control, thereby weakening the Europeans relative to the United States. Secondly, unless physical possession is taken of the investment certificates, which is not common for governments, these assets are subject to a freeze such as the one imposed on Iran. Third, recycling the surplus to the Third World is performed by Western institutions which not only control these surpluses but also derive much of the recognition and returns from this activity. Fourth, the nominal value of the surplus has not increased in step with the weighted index of price increases in the bill of goods imported by Arab countries. Inflation and currency depreciation have significantly reduced the purchasing power of these surpluses. Fifth, as Keynes has so aptly put it, 'if you owe your bank a hundred pounds you have a problem, but if you owe a million, it has'. Arab surplus funds are now too large and too narrowly spread geographically and asset-wise. This raises serious questions as to who would suffer more — the West or the Arabs — from major shifts in the value and/or the composition of these assets.

Oil Power: Myth or Reality?

Using the conventional definition of power as the ability to affect decisions of foreign governments, oil has provided the strength by which Arab governments can exercise influence over the policies of other governments at any particular time. However, economic power can be viewed as a defensive weapon rather than an offensive one.

The scope of oil power should not lead us to disregard the risks and limitations inherent in using this tool. Oil has certain characteristics that give it political influence, but it should also be viewed as a commercial commodity in normal circumstances. The strong relationship between oil and politics may be but a temporary linkage, that springs from the unusual circumstances prevailing in the Middle East as a result of the continued Israeli occupation of Arab territories. As a defensive weapon, oil is usable in cases where the security and peace of the Arab nations are threatened by foreign powers. Unless in these exceptional circumstances, however, there should be a limited relationship between oil, as a commercial commodity, and politics. Oil should be regarded primarily as an important source of income and financing for development plans that will lead to the continued prosperity of the oil producers.

The inescapable interdependency of today's world produces a picture of countries conscious of the need for mutual co-operation and benefits, regardless of their different ideologies or diverse political and economic systems. The United States and the Soviet Union are the world's greatest rivals, but both seek improved economic and trade relations. It is only natural that economic needs which cannot be met locally are sought externally, and hence arises the dependence of some countries on others. Without co-operation between oil producers and consumers, neither would have been able to achieve their combined or individual goals. In the interdependent world of today, many problems could be avoided if every country respected the wishes, potentialities and aspirations of others. As Professor R.J. Holsti has suggested:

> Degree of need is one variable element in the successful exercise of influence in international politics. Because economic resources are often scarce — but necessary to fulfill national values and aspirations — needs in the modern world are frequently transformed easily into political influence. It is the need for key raw materials by industrial powers that helps explain how 'weak' countries, as measured by military or economic capabilities, can be mobilized for political purposes . . . If influence is to be created out of economic need, the need must be genuine.[15]

The special status that oil now enjoys derives from the 'genuine' need of the industrialized countries for this vital source of energy. The Western world's dependence on Arab oil as a major source of energy has been traced to the following set of factors:

First, oil was comparatively cheap. Coupled with the discovery of enormous crude oil reserves in the Middle East in the late 1930s and early 1940s, in real terms the price of energy fuels declined by approximately 30 percent between 1950 and 1970, which encouraged energy consumption and reduced incentives for industrial investment in energy-saving products and methods of production. Low oil prices also influenced the persistent preference for oil over coal.

Second, the non-communist world's crude oil supplies were controlled by a handful of enormously powerful international oil companies. These Western-owned monopolies, backed by their home governments, succeeded in directing the world oil balance away from the producing nations to the Western industrialized nations. With their complete command over world oil resources, the industrialized countries did not foresee losing control, and this led to complete dependence on oil as a major source of energy for industrialization.

Third, the United States held a position of dominance in the capitalist world during the post-Second World War era. Occupying this powerful position, which its allies had held before the war, the United States and the international oil companies (five of which are American) 'used this opportunity to virtually ram American-controlled oil down the throats of the world to replace coal'.

Fourth, beginning in the mid-1960s, a growing emphasis on environmental considerations, particularly in the United States, gave oil a further competitive edge over coal. Consequently, despite efforts by governments in Western Europe and Japan to subsidize and protect their coal industries, coal production steadily declined and oil accounted for fully two-thirds of the growth in energy consumption.

The October 1973 oil revolution only reinforced this pattern of consumption. Western nations felt the impact of their tremendous dependence on oil, and thought seriously of re-evaluating their use of oil and investigating alternative sources of energy. However, their failure thus far to find an alternative energy source to oil will increase the importance of oil as an international commodity, both in its economic and political dimensions. Japan is still totally dependent on imported oil. The situation in Western Europe is not very different because North Sea oil is likely to meet only a modest portion of the energy requirements of the area. Moreover, although the United States is better situated regarding oil than its partners in the Western alliance, it will still need to import vast quantities of oil. Increased demand and continued dependency on oil confirm the undeniable fact; oil is and will remain a major force in international relations, despite temporary fluctuations in pricing or market demand, such as those exemplified by the European industrial depression in 1980, 1981 and 1982. As stated by Hanns Maull:

The absolute growth of oil imports shows the increasing damage potential, and the relative growth shows the Arab producers' strong leverage *vis-à-vis* Europe, Japan, and — to a lesser extent — the United States.

Since energy is of overwhelming importance for the functioning of all aspects of industrialized economies and societies, the Arab producers' leverage was considerable — unless the consumer economies could adjust to interruptions in supply in such a way as to prevent major disturbances. While they had achieved adjustments at fairly low cost in 1956 and 1967, the situation in 1973 had changed fundamentally.[16]

To put it briefly, oil has assumed a political as well as an economic dimension in today's world. In effect, oil is the key to Arab political influence over others, and the Arabs' understanding of this fact allows them to exert leverage *vis-à-vis* the Western world.

But what would happen to Arab influence if, in the future, oil lost its present dominance? Some observers feel that the development of alternative sources of energy may reduce the importance of oil and, subsequently, the market strength of the oil-producing Arab countries.

As Professor Nazli Choucri has indicated: 'New sources of energy will effectively change the structure of the world petroleum market and invariably interject new issues in international energy transactions.'[15] In the next chapter, we turn our attention to the consideration of alternative energy supplies and the likelihood of their replacing oil as the dominant energy source.

Notes

1. I have inserted this definition to express the political side of petroleum affairs, which are basically economic in nature. I have not come across the same definition elsewhere. Its relevance or irrevelance is, therefore, my responsibility alone.

2. Hamed Rabie, *The Oil Weapon and the Arab-Israeli Struggle* (Arab Institute for Research and Publishing, Beirut, 1974), p. 21.

3. Some of these questions have also been raised by Douglas Feith, 'The Oil Weapon De-Mystified', *Policy Review*, no. 15 (1981), pp. 19–39. Feith raises these questions from a generally negative point of view that is at variance with our study.

4. Hans J. Morgenthau, *Politics Among Nations*, 5th edn (Alfred A. Knopf, New York, 1973), p. 28.

5. *OPEC Annual Report, 1980*, p. 196.

6. Steven E. Plaut, 'OPEC is not a Cartel', *Challenge*, (Nov.–Dec. 1981), p. 19.

7. Feith, 'The Oil Weapon', p. 4.

8. *Saudi Newsletter*, 18 June 1981.

9. World Bank, Summary Report of the Research Project on 'International Migration and Manpower in the Middle East and North Africa' (1981), p. 7.

10. IMF, *Survey* (4 Sept. 1978), and various issues of IMF, *International Financial Statistics*.

11. OECD, *Development Co-operation, 1981, Review*.

12. Ibid.

13. *International Herald Tribune*, 20 Jan. 1982, p. 9.

14. Ibid.

15. K.J. Holsti, *International Politics*, 2nd edn (Prentice-Hall, Englewood Cliffs, N.J., 1972), pp. 240–1.

16. Hanns Maull, 'Oil and Influence: The Oil Weapon Examined', *Adelphi Papers, No. 117* (The International Institute of Strategic Studies, 1975), p. 3.

17. Nazli Choucri, *International Politics of Energy Interdependence* (Lexington Books, Lexington, Mass., 1976), p. 21.

17 OIL AND ALTERNATIVE ENERGY SOURCES

In modern industrial societies, cheap energy has supported a complex system of production and distribution that lies at the very foundation of economic and cultural progress.

Progess on many fronts is highly correlated with increased productivity. Without a surplus above basic needs, no society can enjoy cultural amenities or invest in technological change, and the size of the surplus is one of the primary determinants of the extent of cultural development and of economic and technological progress.

Productivity itself is dependent on energy. The continuous decline in the price of oil throughout the 1950s and 1960s was responsible for increased capital usage and the substitution of machines embodying new technological advances for ordinary labor. The increased reliance on machines and their associated technologies raised the overall productivity of the economic system and generated the surplus that supported higher levels of consumption of cultural goods, and laid the grounds for further growth and prosperity through research into technological change. It is no wonder that, with the rise in the price of oil, a major brake was applied to this general sequence of development and progress. The increase in oil prices is symptomatic of a basic scarcity in this non-renewable source of energy, and its limited or costly availability is bound to hamper world progress. Immediate responses to this situation are necessary; alternative energy sources must be found, conservation must be encouraged, and rational use must be practised.

Some of the most thoughtful words on the causes of the energy crisis and its dimensions were spoken by His Majesty King Khalid ibn Abdulaziz in an interview with the editor of the Kuwaiti newspaper *Al-Siyassa*. He summed up the determinants of the crisis in the following manner:

(1) Increased consumption caused mounting pressures on oil supplies, which were already reduced substantially as a result of political developments in some oil-producing countries.
(2) The advanced industrialized countries failed to find alternative energy sources for oil, which is being depleted rapidly with its profligate use as a source of energy at a time when modern science is providing more valuable and beneficial uses for oil.

183

(3) High prices were imposed by some major oil companies to reap windfall profits at the expense of both producers and consumers, a course of action attributed to the traditional monopoly of these large companies in existing world oil markets.

These facts point to the inescapable conclusion that the only way available to mitigate the present energy crisis and the increased rate of depletion of known oil resources lies in the search for new, alternative sources of energy that may in time replace oil as the principal energy source. To begin with, there are a number of substitute policies through which global energy demand may be reduced in the short, medium or long terms, including: (i) energy conservation rationalization; and (ii) increased demand for traditional energy sources such as coal, natural gas and hydroelectric power. On the supply side, solutions lie in the direction of: (i) the discovery of new oil reserves, adding to the already proven quantities estimated at 645 billion barrels; and (ii) the achievement of sudden technological progress in developing alternative energy sources, especially in the field of nuclear energy.

The first demand-side principle is, undoubtedly, the former in reducing the extent of dependence on oil, considering that each consumer country has in place or could rapidly develop its own programmes for reducing dependence on imported oil. The other principles, however, have a much longer gestation period and would involve extremely costly solutions. The special programs launched by the Western countries with the aim of conserving energy, particularly the programs launched in the United States, have so far reduced oil consumption by only small degrees, amounting to a miniscule proportion of global oil demand. Between 1973 and 1976 an estimated reduction of only 2.7 percent of total consumption in the industrialized countries was attributed to conservation, while the United States failed to achieve any serious reduction whatsoever. Although total energy consumption in 1975 was reduced by 14.4 percent in comparison with the 1973 level, the major cause of the reduction was a combination of fuel economizing and mild winters in both 1974 and 1975; it was not due essentially to effective energy conservation policies.[1] Conservation policies, involving measures such as imposing speed limits on cars and reducing heating temperatures at home, in general only touched the surface of the problem. In addition, the measures did not require serious economic or technological steps to conserve energy. Even though the demand for oil in most of the industrialized countries had in the past few years declined substantially, it is still unclear to what extent this decline is a response to conservation measures. The latest figures ascribe less than 20 percent of the reduction to conservation measures, whereas a full 60 percent is attributed to depressed economic activity in the West and another 20 percent to the drawing down of inventories.

On the supply side, the discovery of vast new oil reserves has had no great success. There is much talk about developing new sources of oil in 'politically secure and neutral areas' such as the North Sea. There has been increased interest in searching for oil in the waters of the Arctic, the seas of Latin America, South and Central Africa, Australia and Java, the United States, the Gulf of Mexico, and even the Atlantic. But no substantial new reserves have been found. Some estimates put American potential reserves at 440 billion barrels of oil and 2,000 billion cubic feet of natural gas. Such reserves are said to lie in offshore areas, and both estimates exceed the proven reserves of the Middle East.

From time to time, fantastic discoveries are announced in distant and exotic locations, and frequently they are trumpeted by the media, but there is no real proof of the existence of such reserves, and most of these announcements are no more than speculation. Indeed, in our opinion, some announcements may be part of the propaganda campaign launched against Arab and Third World oil production in an attempt to weaken their economic status, shake their self-confidence and disrupt the unity of OPEC.

Notwithstanding the current aberrations in the oil market, increased dependence on oil by the industrialized countries is bound to create mounting interest in the Arab world. The need for oil and the relatively cheap costs of oil production in the Arab countries, coupled with the growing realization that no substantial new oil discoveries outside the Arab region are likely to be found, will undoubtedly emphasize the special economic characteristics of Arab oil.

The deputy director for oil exploration of British Petroleum has expressed his belief that a new Middle East is unlikely to be found, because most of the globe has been initially explored for oil except for parts of Siberia, the Arctic and Antarctic. The world will have to depend on the reserves available in the OPEC countries in general, and the Arab producers in particular. This dependence will last for some time to come.

Again on the supply side, most of the alternatives to oil are still at the experimental stage, although they may prove decisive as sources of energy in the distant future. They include sources ranging from solar energy to wind, but the most important are coal, natural gas and nuclear power. A brief survey of these alternatives reveals the following facts:

Coal

Coal is regarded as a prime energy source, with reserves totalling more than ten times the level of proven oil reserves. Many energy experts consider coal to be the nearest available replacement for oil, unless there is an unexpected technological breakthrough in nuclear power. This is particularly important, considering the vast reserves of coal in the energy-consuming countries. The possibility of reverting to coal as a replacement for oil applies only to those countries that have sufficient technological expertise to use, convert or employ processes – either known or yet to be discovered – that permit the use of coal as a reserve energy source when oil and natural gas are either depleted or greatly reduced. Experts estimate coal reserves at 8,960 billion metric tonnes. Additional vast quantities may be discovered in non-consuming countries which could put global coal reserves at 15,000 billion metric tonnes, although half of these reserves are exploitable under present conditions (see Table 17.1).

On the other hand, some experts doubt whether coal markets can expand in the near future. Others think that coal may lose even its present share of the energy market. These assessments notwithstanding, there are a number of problems in developing coal as an alternative to oil. Some fall into the category of environmental restrictions, which hamper efforts designed to develop coal-mining and foster the greater exploration of coal. Environmentalists, especially in the United States, are mounting wide-scale campaigns against any such efforts. Other restrictions relate to the high cost of coal mining, transport, and other factors that tend to make some coal-mining operations uneconomical. Overcoming these limitations is a matter of great importance to those who advocate expanded coal operations, but major breakthroughs are not expected for years to come.

Natural Gas

Gas, as an important energy source, meets a large part of energy demand

Table 17.1: Estimated Global Coal Reserves (billion tons)

Country	Proven	Estimated	Total	% of reserves
Soviet Union	250	5,377	5,627	63.0
USA	81	1,425	1,506	17.0
China	75	936	1,011	11.0
UK	127	28	155	1.7
India	13	93	106	1.2
South Africa	37	35	72	0.8
Canada	43	18	61	0.7
Australia	49	47	96	1.0
West Germany	62	—	62	0.7
Others	—	—	264	2.9
Total	737	7,959	8,960	100.0

Source: *Alam al-Naft* (Oil World), 28 July 1973.

on a global scale. In the United States — the largest consumer of natural gas in the world — this energy source provided 27 percent of total energy requirements in 1980, or the equivalent of 492 million tonnes of oil. In 1990, however, consumption of natural gas is expected to double to 46.7 trillion cubic feet.

Natural gas reserves are continually adjusted upwards (see Table 17.2) but are still low compared to those of oil and coal. The main problem facing increasing dependence on natural gas lies in its transport to consumers at reasonable costs. Until the Second World War, the high cost of transporting gas limited its consumption to areas close to producing fields. During the last two decades, advances in transport techniques and better intercontinental pipeline networks have made gas available to other areas. The 1960s witnessed the construction of several gas pipeline networks, such as the one between Iran and the Soviet Union. Technological progress made gas liquefaction possible, and it became easy to transport gas in special tankers for consumption overseas. New technologies and advanced pipeline construction opened new opportunities for gas producers in the Middle East and the Gulf countries. Natural gas, which used to be flared or re-injected into oil-wells without any practical benefits, became exportable to Western consuming markets, thereby adding to the income of the countries concerned.

The United States, the Soviet Union, the Netherlands and Canada are the four major producers of natural gas, accounting for 75 percent of total world gas production. In 1980 the United States produced about 37 percent of global gas production, a figure markedly lower than the 49 percent produced in 1973. But while gas production in the

Table 17.2: Arab and Global Proven Gas Reserves (trillion cubic feet)

Country	1973	1974	1975	1976	1977	1981
			Year			
USA	247.3	250.0	215.0	220.0	210.0	198.0
Canada	50.3	52.5	53.4	56.0	58.0	89.9
Soviet Union	706.0	812.0	800.0	918.0	920.0	1,160.0
China	20.0	25.0	25.0	25.0	25.0	24.4
Holland	92.0	94.8	70.0	61.9	60.0	55.7
UK	50.0	50.0	50.0	30.0	39.0	26.0
Norway	33.0	24.7	25.0	18.5	20.0	49.4
Australia	37.7	38.0	32.5	32.3	32.0	18.7
Mexico	11.0	15.0	12.0	12.0	30.0	73.4
Pakistan	10.1	16.0	16.4	15.8	15.5	16.4
OPEC's Arab states	378.5	445.0	438.3	453.2	615.3	203.4
Venezuela	42.0	43.0	42.0	40.7	41.0	47.0
Ecuador	5.0	5.0	5.0	12.0	5.0	4.3
Nigeria	40.0	45.0	44.3	44.0	43.0	40.5
Indonesia	15.0	15.0	15.0	24.0	24.0	27.4
Iran	270.0	330.0	329.5	330.0	500.0	484.0
Arab states						
Total	280.2	421.6	366.5	364.5	379.4	432.1
of which: UAE	13.5	21.5	21.5	21.5	21.5	23.3
Abu Dhabi	12.5	20.0	20.0	20.0	20.0	19.5
Dubai	1.0	1.5	1.5	1.5	1.5	0.8
Sharja	1.5	1.5	1.5	1.0	−	3.0
Iraq	22.0	27.5	27.0	27.0	28.0	27.3
Kuwait	36.5	35.8	35.6	34.2	34.0	30.5
Saudi Arabia	54.9	58.5	106.8	86.0^a	87.5	114.0
Oman	2.0	2.1	2.0	2.0	2.0	2.7
Qatar	8.0	8.0	7.5	27.5	40.0	60.0
Bahrain	4.0	6.6	5.5	3.0	3.0	8.6
Syria	0.7	0.7	1.2	1.2	3.1	3.2
Algeria	105.9	229.0	126.0	125.8	125.0	130.9
Egypt	4.2	3.5	4.0	2.8	3.2	3.0
Libya	27.0	26.5	26.3	25.8	25.7	23.2
Tunisia	1.5	1.5	1.5	6.6	6.4	5.4
Morocco	−	−	−	−	−	0.3
Other states	127.3	310.5	128.0	117.2	125.4	200.3
Total	2,033.4	2.555.1	2,232.1	2,302.8	2,519.7	2,755.4

Notes: a. Saudi Arabia's Ministry of Petroleum & Mineral Resources corrected reserves to 86 trillion cu/ft.
Source: *Oil & Gas Journal*, 23 Dec. 1975; 27 Dec. 1977; and 28 Dec. 1981.
Selected Studies of Oil Industries, issued by OAPEC (Kuwait, 1979).

United States has declined as a proportion of the world total, the Soviet Union's share has risen and now stands at 29 percent, although this is somewhat less than the 1976 share of 34 percent.

Taking production levels and reserves of natural gas into consideration, in 1973 it was estimated that gas would last for 42 years. Adjusted estimates available in 1980 suggest that reserves may last for 51 years. If countries of the communist bloc were excluded, natural gas reserves declared available in 1973 would last for 35 years; the 1980 estimates suggest they would last for 45 years.

Nuclear Energy

Nuclear energy, for some, is the industrialized world's only route to a prosperous future, due to its limitless output relative to its dependence on resource inputs. However, nuclear energy is still largely an infant industry and is as yet far removed from posing any serious threat to oil or other traditional sources of energy. Its overall contribution to world energy consumption does not exceed 1 percent, and it is mainly used for electricity generation.[2] A former director of planning and analysis at the American Nuclear Agency has this to say on the subject:

From several different perspectives, the present can be seen as a watershed in the development of the nuclear industry. For example, the amount of energy produced in the United States from commercial nuclear power plants does not yet equal that produced by burning wood. From this perspective, the nuclear business obviously must be classed as an infant industry. The infant, however, is a healthy one . . . A decade from now, Americans in all sections of the country will be drawing a significant share of their electrical energy from nuclear power plants.[3]

Table 17.3 shows significant growth in nuclear power; in 1973 there were 107 nuclear reactors producing just over 40,000 megawatts, while in 1980 there were 253 reactors producing just under 136,000 megawatts. This indicates that nuclear power stations are increasing worldwide, regardless of the objections raised by environmentalists, and it is likely that nuclear energy will acquire a more important place in the next few years as a significant energy source.

There are many other forms of energy sources, such as solar power, which has its special characteristics as clean, renewable and limitless. Geothermal energy is already exploited in the Soviet Union, France, Italy, Iceland and other countries, particularly for heating. Wind is another potential energy source. All contribute to a potentially prosperous future for mankind, although most are considered to be at the threshold of wide-scale use. For example, solar energy, one of the most

17.3: Nuclear Energy: Installed Capacity, 1973 and 1980

	1973		1980	
	Number	Capacity (MW)	Number	Capacity (MW)
Argentina	—	—	1	335
Belgium	—	—	3	1,664
Bulgaria	—	—	2	816
Canada	6	2,512	11	5,494
Czechoslovakia	—	—	2	800
Federal Republic of Germany	6	2,143	14	8,606
Finland	—	—	3	1,740
France	10	2,818	23	15,409
GDR	2	510	5	1,694
India	—	—	4	809
Italy	—	—	4	1,382
Japan	5	1,749	24	14,994
Netherlands	—	—	2	498
Pakistan	—	—	1	125
South Korea	—	—	1	564
Spain	3	1,073	3	1,073
Sweden	1	440	8	5,515
Switzerland	3	1,006	4	1,940
Taiwan	—	—	2	1,208
UK	28	5,330	33	6,980
US	34	19,776	70	51,550
USSR	9	2,775	33	12,616
Total	107	40,112	253	135,812

Source: Original information, plus the 'Oil and Energy Trends', *Statistical Review*.

promising future sources, suffers from a major drawback — it is difficult to store, and as the sun and earth rotate, it is not available on a steady basis. Again, although the inner layers of the globe still embody areas with high temperatures, geothermal sources are generally expensive and uncertain.

Less exotic but equally difficult to extract is the oil tar-sands and shales. Not only does the process of oil extraction from these deposits involve major environmental erosion, but the processes are only economic at very high prices. For this reason, several projects were shelved in Canada in 1982 as soon as the price of oil showed the first signs of softening.

Several conclusions may be drawn from the above arguments.

(1) The demand for energy, especially oil, is not likely to decline significantly up to 1985.

(2) The development of new alternative energy sources to oil is still largely at a theoretical stage. Such alternatives are unlikely to account

for any substantial portion of energy requirements in the foreseeable future. Some new development projects, such as shale, currently appear uneconomical, although available in vast quantities. Other projects have been shelved or totally ruled out for the present. Nevertheless, some alternative sources can be developed to become limitless energy sources, pending the availability of the requisite technological knowledge and the practical application of methods that have already passed through the experimental stages of development. Solar energy and atomic fission are cases in point. When the production of energy from these sources becomes economical, they will undoubtedly have far-reaching effects on the pattern of energy consumption and, consequently, on oil consumption.

(3) There is little indication from the OPEC countries of any substantial new proven quantities of oil, despite the illusory discoveries announced from time to time in the media, with the apparent intention of confusing OPEC members and shaking their unity as an organization.

The present trends in energy consumption indicate increased dependence on oil, notably the oil produced by the Arab countries, which control more than 52 percent of the proven global oil reserves. Oil will remain for the present the prime source of energy and will continue to play an important role in the international balance of energy. As for the consuming countries, energy conservation today is the only practical course for reducing the global demand on energy.[4]

Notes

1. OAPEC, *Oil and Alternative Energy Resources* (Kuwait, 1977), p. 28 (in Arabic).

2. Robert H. Connery and Robert S. Gilmour (eds.), *The National Energy Problems* (Lexington Books, Lexington, Mass., 1974), p. 54.

3. Ibid., p. 63.

4. Henry Kissinger, 'Energy: The Necessity of Decision', *The Atlantic Community Quarterly*, vol. 13 (Spring 1975), pp. 11-12.

18 CONCLUSION

The oil industry, oil markets and oil policy clearly involve complex issues of economics. However, they are by no means purely economic questions and to view them in terms only of economics and commercial interests is to adopt a distorting and misleading viewpoint. Oil in the Middle East belongs to governments. Decisions concerning the pace of exploration and development and the rate of extraction are political as well as economic decisions. The regulation of oil flows among nations is viewed by all governments primarily as a political function. It can be both the objective of policy and the tool by which policy objectives are pursued. Security, equity and improved international relations are the political concerns of governments. Our subject is not the economics of oil, but rather the political economy of oil.

The political dimensions of oil are nowhere more evident than in the case of Arab oil. The main thesis of this book is that the oil crisis is rooted in the Palestinian question rather than in the seeming imbalances between the supply of and demand for oil. Oil has always been a major factor in most of the political crises of the Arab area, although it did not step into the center stage of events until recently. Much of what has happened in the Arab area since the discovery of oil is related directly or indirectly to the Palestian question. If oil did not appear in the 1940s and 1950s to be linked to political events in the area, this must be attributed to the full control and manipulation of the oil industry by the Majors and their mother countries. The Majors determined the economic aspects of oil production, pricing and exploration, and the Western owner countries determined the political framework of its exploitation and distribution.

Arab governments and oil experts were aware of the hegemony of the oil companies and their imperialist owners on all oil matters. Nonetheless, they saw the potential power, embedded in the control and use of oil, to alter the world politico-economic order in a way that might improve the position of the Third World. But the Arab governments then lacked the experience and tools to take the initiative in repatriating the oil industry and bringing about this necessary change. On the other hand, the oil companies and the Western countries acted cohesively, perhaps not according to a preconceived plan, but surely in a manner that exploited the prevailing advantageous conditions

192

in the oil industry. One thing is clear: the Western governments acted decisively and unabashedly to promote their national interests and the interests of their national stockholders in the major oil companies, whereas the governments of the oil-producing countries were either unable to use, or unconscious of, their potential power.

The end of the colonial era and the emergence of national liberation movements in the 1950s, the attainment of independence by many Third World countries in the 1960s, and the decline of Western influence and power in the Middle East combined to trim the influence and role of the multinationals in the oil industry. The establishment of OPEC and OAPEC, and the triumph of the producing countries in their negotiations with the multinationals during the period 1970-3, which culminated in the oil revolution following the Ramadan war of 1973, paved the way for the ultimate reorganization of the industry, and culminated in the oil producers gaining control over their main natural resource.

The association of the triumph of the oil-producing countries with the Ramadan war explains the nature and extent of the relationship between oil developments and the Palestine question. The Arabs' preoccupation with the Palestinian question prompted them to politicize oil in the first place, and it was the realization of oil power that finally brought about the reversal of the hierarchy of relationships between the oil companies and oil producers. Thus the Palestinian question was responsible for the oil revolution.

The planting of Israel in the heart of Arab land had and will continue to have enormous implications, not only for the immediate region but for the whole world. Its impact on the oil industry is only one aspect, albeit a major one. The Arabs' resort to the oil weapon has been defensive and in response to Israeli expansionism and aggression. It is a non-violent instrument that cannot be divorced from the context within which it operates. Its successful use in 1973 was the outcome of the experience gained from the ineffectual attempts in the previous Arab-Israeli wars of 1948 and 1956. It is now axiomatic that, as long as the Arab nation is threatened, oil will be used in its defense. The main threat to the Arab nation is Israel, and for this reason the Palestinian question will remain the most important security issue for the Arabs and the world. If there is an energy crisis, it is because there is a Palestinian crisis. Future interruptions of oil supplies remain a distinct possibility so long as the Palestinian problem remains unresolved.

Before 1973 oil experts were not sure that Arab oil could be used effectively to influence Western positions towards Israel. Previous

Arab failures to present a united front were often cited as the main causes for their ineffective use of the oil weapon before 1973. Thus the oil revolution of 1973 came as a shock to the West, despite repeated Arab warnings and the historical changes in the oil market that enabled the Arabs to make oil power effective. It took almost a century of struggle for the Arabs to attain their ultimate objective of control over their own oil. The history of this struggle is illuminating and is reviewed briefly here by grouping this period into four major sub-periods.

The Initial Period, 1859-1950

During this period, the oil market was under the complete control of the oil cartel led by the Seven Sisters. Together they dominated and determined the production, refining and marketing of oil.

The Oil-Consciousness Period, 1950-1970

Nationalization of major oil installations began and OPEC was established. Although few major developments occurred during this period, it laid the foundations for the major events to come. The establishment of OPEC cemented the growing feeling of unity and fostered collective decision-making among its member countries.

The Take-off Period, 1970-1973

This is the period during which the balance of power changed in the oil industry. Oil producers gained more power and confidence in their ability to realize their common objectives, whereas the major oil companies lost influence and control. The Arab countries of the Gulf and Libya played a major role in this transformation process, but its implications and momentum reverberated in every oil-producing country. The three decisive meetings in Tripoli, in Tehran, and once again in Tripoli rewrote the power equation in the oil industry for all time.

The Oil Revolution, 1973

The Ramadan war was the straw that broke the camel's back. The shock-wave of the war and the successful use of the Arab oil weapon finally toppled Western domination of the oil industry. Oil producers came to assume full control over production, pricing and exploration. The Saudi Arabian role proved to be decisive in precipitating this long-awaited change, bringing about national sovereignty over natural resources and dismantling the old, unjust economic order.

Joint action and solidarity assured the success of the Arabs' use of the oil weapon. On the military front, Syria, Egypt and Jordan co-ordinated their campaigns efficiently and effectively. At the same time, the Arabian Gulf countries and those of North Africa provided strategic depth for the confrontation states and marshaled their financial and oil resources in support of the war effort. Saudi Arabia played a pivotal role in activating Arab oil into a devastatingly effective political weapon.

A brief review of the history of these recent developments is in order here. The politicization of Arab oil was a natural evolutionary response to the historic imperialist designs, to carve the Arab area into zones of influence. The Red Line Agreement of 1928 was a clear example of the collusion of the oil cartel to allocate the sources of Arab oil among its members. The implantation of Israel in the Arab world is inseparable from these early designs and a natural and logical extension of the cartel's colonialist imperatives. Israel represents 'the only strategic reserve left for the United States' in the Arab area, according to President Reagan. It is therefore not surprising that Arab–American relationships have remained and continue to be hostage to Israeli demands on Washington. The Arabs were frustrated in their repeated attempts, by peaceful means, to improve their relationships with the West and to neutralize its flagrant bias towards Israel. They found themselves obliged ultimately to resort to the oil weapon to awaken the West to its vital interests in accommodating the Arabs and in finding a just and lasting peace in Palestine. The change in Western attitudes, particularly those of Europe, towards the question of Palestine has been the direct result of Arab oil pressure. A major schism has developed between the United States and Europe on this issue. Although the rift between Europe and the United States is not solely the outcome of their disagreement on how to deal with Israel, the divergence in their Middle East policies has exacerbated their other differences and widened the rift.

The failure of the United States to provide Europe with its oil requirements (American companies actually diverted oil shipments from Europe to the United States during the 1973 embargo) deepened

European dependence on OPEC oil. European countries rushed to secure their oil requirements through bilateral agreements with Arab oil producers, despite loud protestations by the United States, which preferred a joint Western response. The Americans feared that bilateral deals with oil producers would eventually break the ranks of the Western alliance and give Arab producers undue influence and power. In the end, even the United States succumbed to the oil shortages and initiated a number of moves designed to 'balance' its Middle East policy. The Ramadan war was perceived by the Americans as a golden opportunity to restore United States influence in the area.

The normalization of Arab-American relations was also Egypt's objective after Nasser's death, and the intersection of the interests of Arabs and Americans in restoring bilateral relations opened new vistas for improvement in Arab-American relations. The first improvement came immediately in November 1973, following Kissinger's visit to Egypt.

Meanwhile, Soviet-American relations deteriorated during the Ramadan war. However, Soviet interest in maintaining the world strategic balance gave the Americans a significant advantage in implementing undisturbed their 'peace initiative' in the Middle East. Sadat was lured into the Camp David Agreement, which effectively fragmented the Arab front and gave Israel unequivocal military superiority on the Eastern front. The Fahd Plan and the recent Reagan Plan are new attempts to break the deadlocked Camp David Accords. A new framework is needed for a comprehensive settlement of the Arab-Israeli conflict, but nothing short of an independent Palestinian state, with East Jerusalem as its capital, is acceptable to the Arabs. Both the Fahd Plan and the derivative Arab plan that gained consensus at the Fez Summit Conference in 1982 sketch the absolute minimum acceptable conditions for peace in the Middle East.

The Arab conditions for peace are backed by the full power of oil and finance. But the effectiveness of either of these instruments is contingent on the prevailing market conditions at the time of use. In general, the greater the dependence of the consuming countries on oil imports, the greater their vulnerability to the oil weapon, and the more significant the leverage the Arab oil producers have on them. Given the limited availability of economic alternatives to oil as a source of energy and the oil-intensive economies of the West, Arab oil will remain a potent political instrument in the international arena, notwithstanding temporary aberrations in the world oil market of the type being experienced now.

Actual power and influence, however, are not synonymous with potential strength and command over strategic resources. Effective use and exploration of potential strength is still the major determinant of a country's international influence and stature. Proper exploitation of potential strength does not provoke undesirable responses and results in net gains. The efficacy of the use of Arab oil as a political instrument should therefore be judged in terms of the net gains that have been generated or are likely to be generated by its use.

Although it is difficult to assess all the benefits and costs associated with the use of oil as a political instrument, it is clear that it has focused the attention of the West on the need to accommodate Arab oil-producers, who now appear committed and able to use oil effectively as a non-violent instrument to realize Arab national objectives and force an acceptable resolution of the Palestinian problem.

Oil and petrodollars, effective as they may be in substantiating Arab positions, are no substitute for military effectiveness. Economic power must be coupled with military power, for neither alone is sufficient. International influence is, and will continue to be, a function of both dimensions. The relative importance of economic capabilities for Arab international influence is, however, decidedly superior to military power. The Arabs' military power is still not commensurate with their economic power. However, there are also some novel international dimensions that must be taken into account. First, the resort to military force as a means of imposing political solutions is no longer feasible. Gunboat diplomacy is dead. As Klaus Knorr has correctly observed, threats backed by military force will not scare any Arab government into changing its policy, and a small-scale military campaign to invade and replace local governments with collaborative ones will most likely fail. To occupy and secure any Arab oil-producing country, even the very small ones, would take a long period of time and would be a difficult and uncertain task.[1] Second, with the exception of the United States and the Soviet Union, no other industrialized country commands the military strength to undertake large military operations overseas of the type required to occupy the oilfields. Even the United States is still in the initial stages of preparing a Rapid Deployment Force. Third World public opinion is no longer docile. It is now an active force that precludes international adventures and the imposition by force of the will of the strong on the ill-prepared. Furthermore, domestic public opinion in the countries most capable of undertaking military campaigns overseas is generally hostile to military adventures. The disillusionment of the American people with the Vietnam war is the best available guarantee against future American armed intervention in the Third World. Fourth, the possibility of nuclear conflict between the superpowers over differences in the Middle East has so far imposed strict discipline and self-restraint on both of them. This has helped militate against major military campaigns by either power in the area. The Soviet Union has concluded a number of treaties with several Arab governments that bind it to support these governments if they are attacked. This suggests that any conflict in the Middle East is bound to spread uncontrollably to engulf the whole world.

These factors combine to make economic weapons relatively more potent than military prowess. But again, neither is a perfect substitute for the other; if anything, the two are complementary.

Concluding Remarks

Our analysis of the interactions of oil and politics leads us to conclude with several important observations.

(1) The politicization of oil is intertwined with the festering Arab-Israeli conflict. The resolution of this conflict is bound to reduce tensions in the area and will most likely reduce the risks of interruptions in oil supplies. However, this does not necessarily imply that the resolution of the conflict will lead to a major reduction in the price of oil. Indeed, greater weight will be placed in price determination on the finite and non-renewable nature of the resources and lesser weight on its politicization.

Senator William Fulbright has emphasized the direct linkage between the Middle East conflict and current economic difficulties, and speculated that resolving the Middle East conflict would be a sure way to ease the political and economic world crisis. He has also maintained that an Israeli withdrawal from the occupied territories will lead to a major reduction in oil prices which may give the West and the rest of the world the critically needed time to adjust and adapt to the new energy situation.[2]

(2) To the Arabs, the Arab-Israeli conflict derives from the non-resolution of the Palestine problem. The fundamental issue of Palestine has to be addressed adequately before the Arab-Israeli conflict can be resolved.[3]

(3) The participation of the Soviet Union in any future peace

negotiations in the Middle East is necessary and vital. The Soviet Union is a superpower with extensive strategic interests, influence, and presence in the area. No enduring settlement can be reached in the region without the approval of both superpowers. Recognizing these dimensions of Soviet presence and influence on the Middle East has prompted Senator Kennedy into calling for the participation of the Soviet Union in the search process for peace in the Middle East. He has emphasized, however, that the Soviet Union must demonstrate at every step its continued commitment to the process and its support for the security of both Israel and its Arab neighbors.[4]

The Soviet Union is currently excluded from the process, given that the United States is still the only power acceptable to both sides of the conflict. The Soviet invasion of Afghanistan in 1979 shattered the Soviet Union's image in the Arab and Islamic world as a supporter of the Third World. The Soviets must demonstrate their interest in world peace in general before they can play an acceptable role as a peacemaker in the Middle East. The unwillingness of some Arab states to accept the role of the Soviet Union in the process will constrain Soviet diplomacy in the area and will undermine its effective and useful role.

(4) The outbreak of a fifth Arab-Israeli war is highly likely in the event of the failure of the current peace proposals. The outbreak of hostilities in the region carries with it necessarily the possibility of again using oil as an instrument of pressure to support Arab positions.

The successful use of oil during the Ramadan war of October 1973 does not guarantee its successful use in the future, because new conditions call for a re-evaluation of the Arab oil strategy. Besides, many European countries have already softened their support for Israel and now take positions supportive of some Arab demands. Even the United States has recently introduced undeniable changes in its Middle East policy which are more understanding of the Arab position. However, these changes are limited, and are threatened by the pressure and manipulation of the Zionist lobby. Nevertheless, these changes need to be taken into account in the formulation of a new Arab oil strategy for the eighties.

(5) The proposition that the United States can protect its vital interests in the Middle East through Israel is patently naive and wrong. The only sure way for the United States to secure and protect its oil interests in the region is through better relations with the oil countries themselves.

The Arab world is still at the initial stages of industrial development. The huge financial surpluses accumulated by Arab oil-producing countries have yet to be transformed into productive assets, physical or human. In this pursuit, the Arab world has to depend on the industrialized countries of the West. There is a community of interests for both parties. The Arabs need industry and technology, and the West needs oil, finance and markets. There are great opportunities for mutual advantage based on this mutual dependence. Neither party can promote its interests independently of the other party. Together, they can smooth the West's transition to another energy source, while in the meantime the Arab world finds an alternative revenue source to oil and builds a broader economic base. Elizabeth Monroe and Robert Mabro were very perceptive when they remarked that the growth of the world economy at rates comparable to those which prevailed before the oil crisis will ensure that the oil producers' gains are not at the expense of oil consumers.[5]

While the Arab-American relationship is mutually beneficial, the American-Israeli relationship is a one-way dependence, with Israel being a major drain on American resources. The burden of Israel on the United States has been put at over $40 billion since its inception.[6] Israel has received over 50 percent of total United States economic and military aid since 1950. In the fiscal year 1980, each American taxpayer paid over $37.00 directly to Israel. This translates into $1,371 per Israeli citizen, or almost $7,000 for an Israeli family of five. This represents a kind of welfare system unduplicated anywhere in the world, certainly not even in the United States itself. It might interest an American inner city dweller to learn that a larger chunk of his tax dollar went to Israel than to urban renewal! The American environmentalists might be appalled to know that he gave Uncle Sam more money for Israel than for conservation and land management in the United States, or for research and development of alternate sources of energy! The comparisons are endless.[7] It is this situation that prompted Harry Trimborn to declare in the *Los Angeles Times* that Israel is like a person with an artificial lung, whose life preservation is depleting the American Treasury.[8] The same idea was also articulated by John Davis, who stated that America's interest in Israel's existence is based wholly on a commitment and not on need. He goes on to argue that the commitment to Israel has a negative impact on the national and vital interests of the United States in the Arab region, and that it has cost the Americans dearly to maintain Israel's economy and army. Davis's main criticism of this obstinate commitment to Israel is that it may one day result in a clash between the superpowers and the eventual destruction of the world.[9]

(6) The Arab oil revolution of 1973 has successfully dismantled the old and unjust world order. The Arabs' victorious use of oil has opened the door to the emergence of new political balances and hierarchies in which the Third World is no longer on the side-lines of history. In the words of the International Institute for Strategic Studies, the events of 1973 represent a historic happening which is of the same impact and importance as the Chinese revolution.[10]

The real challenge facing the Arabs in the 1980s is to use their oil as an opportunity to promote economic prosperity, and to reaffirm their political sovereignty and international influence to effect legitimate Arab demands for a just and lasting resolution of the Palestine problem.

Notes

1. Klaus Knorr, 'The Limits of Economic and Military Power', *Daedalus* (Fall 1975), p. 234.

2. J. William Fulbright, 'The Clear and Present Danger', *Atlantic Community Quarterly*, vol. 12 (Winter 1974–5), pp. 422–4.

3. Walid Khalidi, 'Thinking the Unthinkable: A Sovereign Palestinian State', *Foreign Affairs*, vol. 56, no. 4 (July 1978), pp. 695–713.

4. Fred W. Neal (ed.), *American Soviet Détente, Peace and National Security* (Fund for Peace and the Centre for the Study of Democratic Institutes, Santa Barbara, Calif., 1976), pp. 19–20.

5. Elizabeth Munroe and Robert Mabro, *Oil Producers and Consumers: Conflict for Co-operation* (American Universities Field Staff, New York, 1974), p. 69.

6. Atif Kubursi, *The Economic Consequences of the Camp David Agreements* (Institute for Palestine Studies, Beirut, 1981), pp. 10–22.

7. Ibid., p. 6.

8. Harry Trimborn, 'Egypt's New Dependence on US Could Give Leverage for Peace', *Los Angeles Times*, 21 March 1976.

9. John H. Davis, 'America's Stake in the Middle East', *The Link*, vol. 9 (Summer 1976), p. 1.

10. International Institute for Strategic Studies, *Strategic Survey, 1973* (International Institute for Strategic Studies, London, 1974), p. 1.

INDEX

Figures and Tables are indicated by (F) and (T) respectively, following the page number.